CHERNOBYL

CHERNOBYL  HEART

20 years on

Adi Roche

Love,
Adi

NEW ISLAND

Chernobyl Heart
First published 2006
by New Island
2 Brookside
Dundrum Road
Dublin 14
www.newisland.ie

ISBN 1 904301 98 3

British Library Cataloguing in Publication Data. A CIP catalogue record for this book is available from the British Library.

Typeset by New Island
Cover design by New Island
Printed in Ireland by ColourBooks

10 9 8 7 6 5 4 3 2 1

The publishers and author would like to thank the following for the kind use of their photographs: Julian Behal (pages 1, 3, 33, 44, 47, 50, 51, 53, 55, 56, 58, 62, 63, 64, 85, 87, 91, 112, 141, 147, 178, 180); Sheila Byrne (13, 88, 148); Sherrie Douglas (161); Paul Fusco (6, 10, 23, 24, 34, 35, 43, 45, 49, 59, 61, 67, 95, 97); Donal Gilligan (22, 30, 31, 71 (top), 94); Sam Gracey (120); Nikki Kahn (133, 177); Anatol Kliashchuk (29, 37, 104, 107, 179); Vasily Lesnighly (83); Maria McCallan (60, 80, 92, 98, 101, 111); Ger McCarthy (42); Ronnie Norton (69); Aodán Ó Conchúr (127, 128); Christine Simpson (103); and Sergey Yatsynov (150).

CONTENTS

'In one country a person costs so much, in another he costs nothing,
and in the third he costs less than anything.'
Jean Jacques Rousseau

'SOS Appeal. For God's sake help us to get the children out.'
Fax from Belarusian and Ukrainian doctors in January 1991

To Belarussians after Chernobyl

As I join the women on the night porch
seesawing their confab back and forth
on rockers, I think of you

or rather how I don't want to think
of what's happened to you as hurricane lamps
fan us with gentle light and gentler darkness

and the background crickets support us with
their singing wings that I can't do justice to,
nor understand, just as I can't the unknown

singing bird in the darkness of unknown trees,
nor the background soprano sopranoing,
except I've always been silently thankful

that I never understood the singer's language,
allowing the words to become wordless music.
And now more than ever I want handicapped

words to turn into such music that will recreate
a miraculous humdrum night such as ours
for you: with voices telling unbelievable tales;

with hurricane lamps, crickets, birds and trees
and one woman watering the blossoming vincas
while the other says its best to water at night;

and how utterly I do not understand this
or anything of what shattered the glass
of your erstwhile ordinary days and nights.

Greg Delanty
From *The Collected Poems of Greg Delanty 1986–2006*

I am delighted to have the opportunity to commend the tremendous achievements of the Chernobyl Children's Project International as we mark the twentieth anniversary of the Chernobyl disaster.

This book contains graphic images of the suffering and pain the people of Chernobyl have endured. It is also a photographic testament to the undaunting and selfless commitment of all those associated with the Chernobyl Children's Project.

We in Ireland have been greatly distressed by the appalling loss of life and the unfolding consequences to human life. Through the work of this Project we are constantly reminded of the humanitarian tragedy and of a catastrophe which has induced such suffering and will continue to do so for many generations to come.

Thankfully this organisation has benefited from the heartfelt support of our nation, which serves to help victims and survivors of the disaster in rebuilding their lives. It is the great sense of love and charity that drives Irish people to give in far greater proportion than our numbers would suggest.

I am personally delighted and heartened at the very positive outcome of recent diplomatic efforts which serve to ensure that the victims and the survivors of the Chernobyl disaster are not forgotten.

I congratulate you and wish you continued success with your future projects.

xi

Mary McAleese
President of Ireland

INTRODUCTION

Standing at the podium in the United Nations General Assembly, I looked out into the auditorium. In the seconds before I spoke, I felt the power and sacredness of this chamber where matters of life and death are regularly discussed. In this hallowed hall I was about to give the most important speech of my life.

The year was 2004. We were gathered to mark the eighteenth anniversary of the Chernobyl disaster under the flags of every nation in the world, in the place where governments pledge to serve and satisfy everyone's legitimate claim as citizens of the world to a sense of hope, a sense that underpins and copper fastens the United Nations Charter, particularly that Convention that protects and promotes the rights of all children. I was there to bring a tragic story to this gathering. I was to testify and seek justice on behalf of those who have survived, those who have died, those yet to be born and those whose tragedy has been relegated to the realms of history – the innocent victims of Chernobyl. This is part of what I said:

My qualifications to speak at the United Nations are simply human. They include my heart, my mind and my history. My history can be spelled with a small 'h' and a large 'H'. My small 'h' lifts a young airline office administrator from a cosy job and home to the world stage of struggle. My large 'H' stands for an Irish woman who grew up learning that it was the fire of struggle that set our nation free to chart its own course; that it was

out of struggle that we survived the Great Famine in the mid-1800s, thanks to the aid and support of generous men and women in many parts of the world, but particularly from the great heart of America. It is this echo that resonates in my very bones and permits me today to speak of the daily struggle of thousands of people, particularly the struggle of the children, who endure the devastating consequences of that fateful day 18 years ago on 26 April 1986.

I cannot speak with the authority of a scientist or doctor, I cannot prove my statements with laboratory or field test experiments, I have no medical or scientific academic qualifications to endorse my remarks. In humility I can only offer my truth, my witness of what I have seen and heard over the past 18 years in my daily work with the victims of this Chernobyl tragedy; to share from my heart my own first-hand experiences of what I have seen and heard; to share what I have learned and been shown, what I have read and studied; to call to mind the human face of that catastrophe, that suffering, that sacrifice; and now that neglect, that negation of reality. Particularly today, I appeal to the world not to forget the innocent children, who have suffered so much because of their particular vulnerabilities – their growing bodies soaking up so much from their radioactive-contaminated world which distorts and destroys their young lives.

This opportunity to speak at the United Nations was the pinnacle of all I had worked for for those past 27 years, the chance to speak the truth and to describe the journey that had brought me to this place.

In early 1977, having recently married Seán Dunne, I began to hear talk of nuclear power 'coming' to Ireland. Both Seán and I were clueless about what nuclear power was. I was suspicious of it without knowing what was involved, while Seán seemed far more positive about it. I began to ask questions and kept haranguing Seán, saying we should find out more. Then the Irish government revealed the proposed plans to build a nuclear power plant in the Carnsore Point area of County Wexford. Immediately there was a huge outcry from the residents of the area and generally from environmental and political activists around the country, whose opposition was mainly based on safety and contamination health issues. I knew then that I wanted to get involved and I eventually persuaded Seán to come along to a meeting of the Cork branch of the environmental group Friends of the Earth, one of the many groups opposed to and campaigning against the construction of the nuclear power plant at Carnsore Point.

Along with over 25,000 other protestors, we attended a huge anti-nuclear rally in August 1978 at Carnsore Point and came back from that marvellous weekend committed to the anti-nuclear cause. We were encouraged and inspired by the many wonderful speakers we heard there, including Petra Kelly, the German Green politician and campaigner, by the songs of Christy Moore, Andy Irvine and Jim Paige, but mainly by the conviction that nuclear power was an energy system to be opposed in all its manifestations.

On 27 March 1979, an accident occurred at the Three Mile Island nuclear plant near Harrisburg, Pennsylvania. My older brother, Donal, lived near the plant and I feared for him and his family. His wife, Leah, was about to have their second baby and we were worried about their situation, especially when we heard that they had to be evacuated. At around the same time, the film *The China Syndrome* was on general release. It told the terrifying story of a nuclear reactor meltdown at a US plant. I was greatly affected by the film and by the danger that the Three Mile Island accident might pose for my brother and his family. This was personal – nuclear power was having a direct impact on my life and my family. These pivotal episodes changed the direction of our lives, especially mine.

The campaign by the Irish Anti-Nuclear Movement against building the plant at Carnsore Point won immense popular support, and eventually the government lost the battle. It not only shelved plans to build the reactors, it also abandoned proposed uranium mining projects throughout the country. With such an early introduction to how well democracy can work and how people can make a difference, Seán and I became involved in the Irish Campaign for Nuclear Disarmament (ICND) with a view to ridding the world of nuclear weapons. This was at the pinnacle of the Cold War era, when the threat of possible nuclear war was ever-present in people's minds as the stockpile of nuclear weapons amongst the superpowers and their respective allies was increasing at a frightening rate.

We became so concerned about the threat of nuclear war that I resigned from my airline position to become a full-time voluntary worker for ICND and we set up an office in our home in Ballincollig, County Cork. In 1982 I devised and implemented a Peace Education Programme for young people in response to their palpable fears of environmental disaster and the threat of nuclear war. The programme gave me a chance to introduce some issues not included in a school curriculum. During the bleakest moments of the 1980s, against a backdrop of the sabre-rattling superpowers of the Soviet Union and the United States and the spiralling nuclear arms race, the spirit, energy and enthusiasm of the young people I was working with helped to carry me through. I continued with this work until that fateful day in January 1991 when I received a fax from Belarus. From then on my life would take a different path.

Looking back and reflecting on what has motivated and sustained me in what could be perceived as despairing work, I think a large part of the answer lies with my family history. A sense of moral responsibility, along with a keen sense of justice, was instilled in all of us. My father was a patriot who loved everything Irish. He set high standards for himself and his family. His drive and belief that anything was possible if one applied oneself and was prepared to do the hard graft to achieve it was paramount in his philosophy of life. He and my mother were active members of the local St Vincent de Paul Society and Meals on Wheels, so my two brothers, sister and I grew up recognising that we had a Christian duty to lend a helping hand. Being a volunteer is in my blood. The understanding from an early age that not everyone is equal gave me a deep sense of responsibility towards people's rights. I don't remember when I first became aware of the term 'volunteer'; all I know is that the early introduction to volunteerism in my formative years shaped the pathway of my journey from then on.

I grew up in a pretty blissful, free and easy way in Clonmel. My childhood best friend was Ann Condon. We lived in each other's pockets from the time we could walk and talk. We went through all the significant events of growing up together – we met and fell in love with young lads together, shared the same desk at school, failed the same exams. We were both outgoing, fun-loving girls and spent more time working on our social lives than we did on studying. All this was shattered when we entered our final year in school and Ann was diagnosed as having leukaemia. We didn't know what it was until we looked up a home medical journal. We learned the word 'radiation' and both of us concluded that her days were numbered.

When we parted and Ann was taken to hospital in Dublin, we wrote stead-fastly to each other. Our letters were always full of jokes and fun, never addressing our fears of death – a death which, when it came, was painful and full of unanswered questions for me. I remember standing by Ann's coffin in the church, walking after it into the graveyard, watching it being lowered and covered by the dark earth. I was angry and kept asking 'Why?' I thought my heart would break and wanted nothing better than to die with her. However, in the years since her death, I have often sensed her encouraging me to go forward with the questions and to seek answers. She and I had learned during her time of illness that this stuff called 'radiation' also came from things called nuclear weapons' tests, and while we knew nothing about what that really meant, we had a deep sense that it was bad. We both had heard in school about the Hiroshima bomb and weapons being tested in the air and how radiation leaked from Windscale/

Sellafield and wondered if that could have affected her health.

In the intervening years, I have often sensed Ann's presence and protection in my life. I even think she had a hand in choosing my life partner, Seán. I remember attending Ann's sister Nuala's wedding and being at the church in the small village of Newcastle, County Tipperary, and seeing Seán for the first time and deciding there and then that he was the one. It was uncanny, really, but where I saw him first was in the church-yard within feet of where Ann was buried. Coincidentally, Seán subsequently told me that he had been the one to bring the news of Ann's death to his friend Brendan, who later married Nuala. So while Ann was no longer alive, I feel she somehow replaced herself by bringing Seán into my life, who has become my most loyal and trusted friend and partner.

I hope that with my life's work I have found some of the answers to the questions that Ann and I had and that in so doing I have contributed, in some small way, to ensuring that others do not suffer as she did.

A PERSONAL JOURNEY

'The splitting of the atom has changed everything except our way of thinking, and thus we drift towards unparalleled catastrophe.'
Albert Einstein, 24 May 1946

When the accident at Chernobyl happened in 1986, Irish CND, in conjunction with the Irish Medical Campaign for the Prevention of Nuclear War, provided a 24-hour information hotline giving basic help and support to the general public, who were deeply concerned about the safety not only of themselves but, more specifically, for the well-being of their children. Our medical team was led by two Cork-based doctors, Mary and Seán Dunphy, who provided useful information regarding simple precautions parents could take to keep their children safe.

After the initial emergency receded, there was little or no information about the scale or consequences of the accident during the following few years. Every so often there was the odd article in some papers, but nothing that caught the imagination or interest of the general public. The word 'Chernobyl' disappeared from practically everyone's consciousness. I continued to include the issue in my Peace Education Programme and tried to learn as much as I could regarding the aftermath of the calamity. It wasn't until January 1991 that I received a fax message which broke through the earlier years of silence and sparse information. I will never forget the message because it was to have a major impact not only on the direction of my work with CND, but on the rest of

my life. The message was simple and to the point: 'SOS Appeal. For God's sake help us to get the children out.' The fax originated from Belarusian and Ukrainian doctors, begging that the children of Chernobyl be taken away from their radioactive environment so that their bodies had some chance of recovery.

The fax was our first breakthrough. Our response was immediate, with the first group of Chernobyl children arriving in Ireland for rest and recuperation in the summer of 1991. The children touched us deeply with their beauty, effervescence, love and friendship. It's one thing to know and understand what radioactivity does to human beings, but meeting, seeing and getting to know and love the child victims of radiation is another story altogether. Up to that point I was able to intellectualise the effects of radiation, and could reel off the facts and figures about nuclear power and nuclear weapons. It wasn't until I started working with the children and their communities that I fully realised the dark tragedy that had unfolded itself into the lives of those affected. What I knew in my head I could now translate into something real in my heart.

The day I received the faxed SOS appeal, I was working with my great friend and colleague Norrie McGregor, and together we started to work on trying to do something for the children of Chernobyl. Eventually both of us, with another great friend, Mary Aherne,

decided that this was where we wanted to focus our work. We eventually formed a new organisation and called it The Chernobyl Children's Project International, an organisation that would work exclusively on the issue of Chernobyl.

Later that year Mary Aherne, Mary Murray and I became the first Irish women to visit the Chernobyl region. Our first trip proved to be a harrowing one. The authorities controlled every aspect of our visit, but we still managed to get an understanding of the situation. I remember visiting a place where children had been abandoned at birth by parents who were unable to cope with their babies' deformities or retardations. Seeing the pain of the babies and children suffering without the help of painkillers and basic medicines was almost unbearable. We saw babies with no mental capacity and babies in permanent pain with distorted and broken bodies. They were oblivious to life around them, locked inside their tormented shapes, awaiting death. We saw children without limbs, a little girl who was so badly contorted that her legs grew up towards her body, youngsters with cruel mental handicaps and others with huge growths on their heads and bodies. The doctors and nurses, who loved their patients, could do little, since they lacked even the most basic medicines and equipment. Mothers and doctors alike made no secret of the fact that if they had had the equipment to identify the foetal

defects in the womb, they would not have allowed these children to have been brought into the world.

In Minsk, Belarus I held a little boy called Uri in my arms and literally felt his life draining away from him. His face and his shrunken body will be one of the memories I will always carry with me. He died shortly after we left. He was suffering from hydrocephaly, a condition where fluid builds up around the brain. All that is required to alleviate the pressure that builds up on the brain is a mere 20-minute surgical procedure, called a shunt, which costs $200. But neither the money nor the simple procedure were available. The overwhelming memory I have of visiting this place in Minsk is the sound of Uri's crying, the doctors' crying and our own.

We visited many hospitals, orphanages and communities and each had a story to tell. Visiting orphanages was the most difficult thing to do. It wasn't that these children weren't loved; their parents were no longer capable of dealing with the scale and impact that Chernobyl had had on their lives. These places were full of little boys and girls bursting to be loved and cuddled. They danced and sang their hearts out in order to entertain these three Irish women who had come bringing toys, food and clothing. We were the first foreigners the children had ever seen. My breaking point was when a little girl in one orphanage tugged the corner of my skirt and drew me to the door, crying 'Mamma, Mamma.' The look of expectation in that child's face was enough to shatter my heart.

The impact of this first visit proved to be a catalyst for action, along with the memories of all the children we had met in the institutions and orphanages, the medics and parents. Conditions in most orphanages were the same – cold, smelly and sparsely furnished. We discovered that in winter they often had little or no heating, despite the severity of their weather. Their diets were also poor nutritionally, so many of the children could not fight off even the mildest infection or cold.

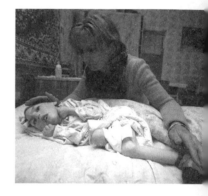

What made it bearable was meeting the children who had been to Ireland and seeing the clear effects of their rest and recuperation in the country. The children who had been with us were faring much better during the severe winter and suffered less illness, since their 'Irish families' had supplied them with multivitamins and tonics.

I remember speaking to Dr Emma in one of the orphanages. She showed us a medicine cabinet which contained only three small bottles of medicine, with no prospect of replacing even those, let alone obtaining sufficient supplies for all the children. There were two radioactive 'hot spots' near the orphanage and the doctor was fearful for the children's health. She herself felt unwell, but said she was afraid to die because then there would be no one to look after the children. Her sorrow was etched in her face. Dr Emma spoke about her personal memory of the accident: 'It was a strange smell, a taste in the mouth and back of the throat, like something metallic was there in the air. I knew nothing about the accident then. I just knew something had changed. It wasn't until I heard it on Freedom Radio [a US-funded radio station transmitted from Western Europe to the former USSR] but I thought it was typical Western propaganda! But in this case it was the truth. I gave a small dose of iodine to my family. I didn't have it for the others in my care. I am sorry, so sorry about that. Many of them now have thyroid cancer. But what else was I to do?'

All we could do was to tell her truthfully that we would do everything in our power to assist them, and from that moment I began to plan a variety of ways to give support. Since that time, I have returned to the zones over 40 times. Each time there has been a purpose, either to deliver aid, bring sick children back to Ireland for treatment, to film documentaries highlighting their plight, to organise building projects or to set up medical health care programmes.

Every time I return to the heart of the Chernobyl zone in Belarus, I witness a country on its knees, struggling to fight against the invisible enemy of radiation, an enemy that is slowly destroying its people. Through the people we have met over the past 20 years, both victims and those working to help the victims, we have seen the true cost of Chernobyl and can only speculate as to what the future holds for the people of the region.

We travelled deep into the radiation zones, delivering aid and talking to the survivors. We measured frighteningly high levels of radioactivity. Scientists working in the area told how the zone is expanding, not contracting as expected, by about 10 km each year, eating into the land around it. One scientist, Professor Vasily Nesterenko, told us that 'it is impossible to say whether we are over the peak of the consequences of radioactive contamination or whether we are just on the threshold.'

What a tragedy. It's a stain on human history. Like so many other countries on the tank paths of World War II, Belarusians suffered desperate atrocities at the hands of the Nazis during the war. They suffered again during Stalin's reign, and then came Chernobyl. Unlike war and its ravages, unlike hunger and disease, radioactive contamination poisoning will never leave Belarus and the other Chernobyl-affected regions.

I wonder what future generations will think about what we have done to this beautiful earth. Unless our descendants come up with a way to neutralise the mountains of nuclear waste and contamination, they will condemn us as irresponsible nuclear joy riders who attempted to ride a wild beast bare backed, halterless and blindfolded.

CHERNOBYL 1986: THE DISASTER

On 16 July 1945, when Robert Oppenheimer and fellow scientists witnessed the first atomic nuclear tests, he realised the destructive nature of the power they had unleashed. Oppenheimer recalled, 'Some of us laughed, some of us cried and some of us fell silent. I was reminded of the words of the Hindu Scriptures: "I am become death, destroyer of worlds."'

This prophetic vision came to pass. It has been 20 years since Chernobyl exploded. The explosions and subsequent fires that wrecked the Chernobyl nuclear reactor brought on what the United Nations has declared to be the greatest environmental disaster in the history of humanity. One hundred and ninety tons of highly radioactive uranium and graphite were expelled into the atmosphere, apart from the radioactive fallout from several tons of nuclear fuel around the station itself. Among all the radioisotopes released, the major contributors were iodine-131, caesium-137 and 134 and strontium-90. Altogether, more than 40 different radionuclides escaped (see the Guide to Radioisotopes on p. 186). Heavier substances came down as fallout, and light products of fission formed as radioactive clouds that were blown northwards and westwards. The lightest substances were lifted into the atmosphere to a height of more than one kilometre. The radioactive plume from the burning reactor moved north and then west over millions of unsuspecting people and on into Poland and Sweden, where it was first detected, taking with it radioactive iodine-131. Caesium and strontium, which are longer-living isotopes, were dispersed over a vast area of 160,000 square kilometres. These radioactive elements also went into global circulation. Fallout was found in practically every country in the northern hemisphere. A rise in background radiation was recorded not only in the nearest neighbouring countries, but as far away as Japan and Brazil. It was as if the reactor had disgorged the entire periodic chemistry table.

The accident occurred as a result of the Soviet authorities' insistence that on 25 April 1986 an experiment was to be carried out at the plant. This experiment went radically wrong at 1.23 a.m. local time on 26 April when a number of fatal mistakes and procedural violations, including the withdrawal of control rods, led to a sudden power surge. Within minutes, the reactor became highly unstable and the control rods were no longer able to balance and control the power surge. Although the station foreman gave the order to go to shutdown, his order came too late. Within four seconds, the power levels ran to 100 times its maximum rating for this kind of reactor, which led to a steam explosion blowing the roof off reactor number 4, which weighed 1,000 tons, as if it was the lid of a saucepan.

Three seconds later, with ruptured cooling system pipes and the control rods blown out, there was another explosion, during which the core of the reactor largely disintegrated. While the operators frantically pumped water into the remains of the shattered shell, it was to little avail because the reactor was now exposed to the air. In the final seconds before the explosion, flames, sparks and chunks of burning red-hot nuclear fuel and graphite went flying into the air. Thundering noises came from all sides of the control room. There was one final colossal explosion, coming from every direction, which made it seem as if everything was bursting apart. A gigantic shock wave, carrying white dust, burst into the control room. The walls and floor of the control room crumbled, debris came crashing down from the ceiling and the lights went out, leaving only three emergency lamps. A flour-like dust invaded the operators' mouths, noses, eyes and ears. Their mouths went dry. People saw the steam escaping that day and by 27 April it had turned to smoke. The ensuing fire burned for 10 full days, consuming a minimum of 10 per cent of the reactor's graphite core.

Heroic efforts were made to smother the fire and bring it under control through flying in and dropping 5,000 tons of clay, sand, lead and dolomite onto the burning exposed reactor. This dramatic and panic-driven action caused the temperature in the remains of the reactor core to rise to 2,000°C, causing the meltdown in the individual fuel channels. On 6 May the fire was eventually extinguished, most likely due to the injection of substantial amounts of liquid nitrogen beneath the reactor. This was done to solidify the soil underneath and prevent a further meltdown from penetrating the foundations of the building, which would have led to the groundwater table, with devastating consequences to millions of people in Ukraine.

It's conservatively estimated that about 100 million curies of radiation were released, although many scientists now believe it was closer to 250 million curies. The day after the fire was brought 'under control', it released 150,000 curies of radiation – more than the total amount emitted during the 1957 Windscale fire in England. By the end of May, more radioactive material was being released each day than the total amount from the 1979 Three Mile Island accident in the United States. In October 1986 the reactor ceased to be a source of airborne fallout when 'the sarcophagus' (a giant tomb-like structure of steel and concrete encasing the exploded reactor) was finally built with 400,000 tons of reinforced concrete.

In Europe, the quantity of caesium-137 deposited by Chernobyl was 400 times more than the peak levels released by all the atmospheric nuclear weapons tested up to 1963. The UN estimates that

8

an area the size of England, Wales and Ireland combined has been contaminated. Of the curies released into the atmosphere, covering 155,000 square kilometres of land, 70 per cent fell onto the population of Belarus. Between the stricken regions of Belarus, western Russia and northern Ukraine, the UN estimates that the fallout has directly and indirectly affected up to 7 million people. The consequences of this fallout will not be fully seen for another 50 years.

In Belarus alone, almost 2.2 million people have been subjected to permanent radioactive contamination of varying density. Of the country's prime farmland and forest, 23 per cent has become contaminated by long-living radionuclides. Thousands of villages and towns have ceased to exist. New spots of radioactive contamination are still being discovered. Large territories in the three affected countries have been declared international ecological disaster zones.

The consequences of the disaster have turned out to be much larger and more serious than many people, including prominent scientists and experts, predicted. Plutonium has spread much farther than the 30-km exclusion zone. In post-mortem examinations, serious traces of radioactivity were found in people who lived several hundred kilometres from Chernobyl. In the Gomel region of Belarus and the Bryansk region of Russia, changes have been found in chromosome

analysis. These changes could have resulted only from exposure to powerful radiation doses of 1,000 rads or more, which is a result of plutonium fission, according to Russian scientist Andrei Vorobyov, director of the All-Union Haematological Scientific Centre. Vorobyov also revealed that published radiation doses were understated.

Owing to the doses of external and internal irradiation received from short-lived radionuclides, the people in the zones have received the highest-known exposure in the history of the atomic age. No people have ever before been continuously exposed to long-lived, man-made radiation. The people of Chernobyl were exposed to radioactivity 90 to 150 times greater than that from the explosion of the Hiroshima atomic bomb. The consequences of Chernobyl cannot be considered in terms of spatial distribution only, since radiation enters the body through the food chain via crops and livestock, wind, forest fires, rivers and streams. The US Academy of Sciences has stated that even low-dose exposure to ionising radiation can cause cancer. The risks are even greater than previously thought and there is no safe level or threshold level of ionising radiation.

The fears for the future of the nuclear industry, which Chernobyl raised in the minds of its advocates, are no justification for the lack of information available to the world community about the true scale of the disaster. The holding back of information was facilitated by the lack of preparedness, resulting in inconsistencies and delays in the action of many Western governments. In West Germany, for example, many restrictions were placed on the movement of children and on the consumption of dairy produce and green leafy vegetables. But in many other countries, people were told that there was nothing to worry about. Much confusion arose over what was an 'acceptable' level of radiation, and this varied from country to country. Even within the environmental movement, little was known about the extent of the damage for at least the first five years. It was as if we had been lulled into a false belief that everything was miraculously all right. Somewhere along the line, we had also been duped by the clever cover-up created by the former Soviet Union. Five years after the accident, I started to research the reason for such a lack of information. I was shocked by what I discovered to be a large part of the answer.

Thanks to the work of a courageous Ukrainian journalist, Alla Yaroshinskaya, I learned that the Politburo (the chief political and executive committee of the Russian communist party) had signed 40 secret Protocols (directives) relating to the suppression of information. I finally understood why I was having such difficulty accessing information. The truth

has been the first casualty and has been hidden until now. Yaroshinskaya, speaking about the significance of the Protocols, claimed, 'The most dangerous isotope to escape from the bleeding mouth of the reactor will never appear on the periodic chemistry scale. It is "lie/'86". A lie as global as the accident itself.'

The Protocols permitted a scandalous censorship and control over the media and thereby stopped the flow of information to the rest of the world. The Protocols controlled the information given to the International Atomic Energy Agency (IAEA) and 'to certain leaders of capitalist countries'. The USSR government commission set up to oversee the accident and clean-up operation destroyed six pages from the official report presented to the IAEA in Vienna. These vital pages held information concerning the amount of radioactivity released in Belarus and the Bryansk region of Russia. The World Health Organization (WHO) had based its figures and findings on the information given to the IAEA, which, I subsequently learned, was monitored and selective. Deliberate misinformation given to the West ensured that the real consequences of the accident remained secret until the mid-1990s. It reminded me of the aftermath of the nuclear bombs dropped on Hiroshima and Nagasaki in 1945 and how the devastating effects were kept a military secret for a decade.

The Soviet Union's long-standing obsessive secrecy about nuclear affairs was despicable and showed an appalling lack of respect for the dangers bound up in the atom. The KGB confiscated files, personal accounts and many documents, denying vital information to science and history. It was a catastrophe born of secrecy. The cold, stark reality of the decisions taken and enshrined in the Protocols has had wide and devastating significance. In Protocol Number 9, for example, the authorities gave themselves permission to increase and change the acceptable levels of radioactivity which people can be exposed to; it was often multiplied from 10 to 50 times. The decision to continually alter these 'acceptable' levels was revealed in Russia's Constitutional Court when the former second-in-command of the

Politburo, Valery Ligachev, told the court that the artificial altering of the limits was for purely economic reasons. If the number of people who had to be evacuated could be kept down, roubles would be saved. As one cynical official said quietly to me, 'It is cheaper to bury them than to evacuate them. You know, if we evacuate, we must compensate and resettle them. Too many problems. It is better that they are left there.'

The suspiciousness and wariness in the people affected by Chernobyl becomes fully understandable when you see how they have been so blatantly lied to and misled. In the aftermath, many herds of cattle were milked and then slaughtered, but instead of destroying the radioactive meat and milk, the authorities made yet another infamous Protocol. Protocol Number 32 reveals that 47,500 tons of radioactive meat was 'mixed' with clean meat and sold to the innocent, un-suspecting population of the Soviet Union, along with 2 million gallons of radioactive milk. This appalling decision was taken to save 1.7 million roubles. The Minister of Health of the USSR recommended that the contaminated meat be distributed as widely as possible in the country and used to make sausages, preserved meat and meat products in a proportion of one to 10 with normal, clean meat. The meat and milk were consumed throughout the entire USSR, thus spreading the danger of radiation to distant people and places. The Soviet Union, always concerned about food shortages, recommended that the contaminated milk be made into cheese and that contaminated grain be given as animal fodder. Years later the Deputy Prosecutor General of the USSR, V. Andreyev, conceded, 'The circumstances mentioned led to contamination of food products with radioactive matter in practically the whole country, and may have a negative influence on the health of the population.'

Economically the cost of Chernobyl is several billion dollars, a cost which is being carried by the survivors and in turn will be inflicted on their children for the foreseeable future. It's a legacy of debt and misery to be passed on from generation to generation, to a people who have already paid too high a price. The disaster has financially crippled Belarus and costs the country 20 per cent of its annual national budget. It has already cost more than 100 billion roubles – over eight annual budgets for Belarus – and it's estimated that the fallout from the disaster will cost Belarus $235 billion, the equivalent of 32 annual budgets for 1985. According to the Institute of Economics of Belarus, their economy will suffer losses of $43.3 billion in the first 30 years after the disaster. Ukrainian economists estimate the economic damage to Ukraine between 1986 and 2015 to be in the region of $201 billion, and in Russia it's estimated that

from 1986 to 1998 it cost their economy $3.8 billion. Because there is no international law governing an accident of this nature, Belarus, Russia and Ukraine have received no compensation for the damage. Near-crisis trends in the economy of the affected countries have prevented proper emergency measures from being taken, such as a completion of the evacuation programme and the importation of clean food to contaminated regions. Compensation and subsistence payments to evacuees are woefully inadequate.

So who was responsible? I had always been an admirer of former President Mikhael Gorbachev of the USSR, primarily because he showed such courage in turning the Cold War around and for his policies of *glasnost* and *perestroika*. These policies eventually led to the end of the Cold War and the break-up of the Soviet Union. Despite the power of the communist party, Gorbachev opened up the Soviet Union to a radical new foreign policy which eventually led to independence for many millions of people. However, I found that his response to the Chernobyl disaster shook my support for his leadership. I got an opportunity to meet and talk with him when he was declared a Freeman of Dublin in December 2001. He told me then that he had deep and serious regrets about how he had handled the aftermath of the accident. Besides being extremely ill informed about just how critical the

Contaminated ships lie resting in a ship graveyard near Chernobyl town

situation was, he felt ashamed that the Soviet administration had miscalculated its response. He said that he defines his life as 'before and after Chernobyl'. While this response gave me some consolation, it did not answer all my questions. Having read in minute detail the exact account of what happened before, during and after the experiment that went so tragically wrong, all I can conclude is that between the deliberate cover-up by the Soviet government, appallingly bad management, a shortage of qualified personnel and a whole series of fatally wrong actions by those in charge at the plant, Chernobyl must now be added to the long list of crimes against the innocents of that region. The accident became one of the key factors in the fall of the Soviet Union five years later. The web of lies surrounding the accident could no longer hide the fallout, both literally and figuratively.

In October 2004 Russia decided to extend the operating life of several of its old Chernobyl-style reactors by at least a further 15 years. Despite pleas from the European Union to close down the 11 Soviet-era reactors, Moscow still insists that it knows best. This is despite experts saying that it's technically impossible to modernise the reactors in question. Moscow's cavalier behaviour is blatantly defying world opinion and is insulting to those who are still struggling to survive the Chernobyl disaster.

The magnitude of the accident has not yet been grasped by the international community. Not only were the reported health effects grossly incorrect, but few are seriously looking at the incalculable damage to the victims. This isn't a calamity where people are dying in the streets of towns and villages. It's a tragedy where thousands of families and communities are quietly suffering, their lives constantly disrupted by ill health and personal crises and corrupted by the nuclear cloud that hangs over them every day.

The people have been subjected to a series of shocks – the accident; the discovery of its true consequences; evacuation; the effects of the break-up of the Soviet Union, followed by the collapse of living standards and the welfare state; and finally, the disintegration into poverty. Besides the effects of the accident, the break-up of the Soviet Union threw the people into disarray. The infrastructure of their lives was community based, and when that collapsed, people were left without the support structures they had previously relied on, such as the local Halls of Culture, Young Pioneers (Soviet youth associations) and collective farms, all of which had played a vital role in sustaining local communities.

This disaster is 20 years old, but its consequences will last forever. Meanwhile, other disasters are vying for the world's attention and Chernobyl has been relegated to history. It's our responsibility to speak out. I'm no longer interested in who is to blame, but I want to unravel what this world of Chernobyl is, to find another sphere with words. I want to look at what we have discovered about ourselves. It's like a mystery that we must solve for ourselves and for humanity. I believe that since Chernobyl, we have been living in a different world. There is a challenge for all of us in that. There is a new reality for us all to face. If we do not know the past, we will not be able to understand the present or make proper decisions for our future. There is no precedent in the history of humanity to which Chernobyl can be compared. In writing this book, I want to share the feelings, the history, the loss, the sensations of a people who have touched something previously unknown. In this book I, too, am a witness, offering a testimony in the hope that my voice will touch a chord in each reader, hoping that it poses questions about the meaning of our lives and our very existence on this earth.

THE FATE OF THE LIQUIDATORS

There is a place known to the survivors as the 'Liquidators' Place'. Its real name is Oksakovshchina and is the Centre for Radiation Research near Minsk in Belarus. In its former Soviet life, it was a luxury retreat for the crème de la crème of the communist party. They chose their hiding place well – it's nestled among the forests, in front of a lake, far from the eyes of the people. Now it serves as a medical and monitoring centre for people throughout Belarus who have been affected by Chernobyl.

On 28 and 29 April 1986, the researchers of the Physics of Reactors Department of the Institute of Atomic Energy at the Academy of Sciences of Belarus made some calculations which showed that the 1,300 to 1,400 kg mixture of uranium, graphite and water released by the Chernobyl accident made up a critical mass for an atomic explosion, capable of producing a force of three to five megatons (50 to 90 times greater than the force of the Hiroshima bomb). An explosion of this magnitude would result in the whole of Europe being exposed to enormous radioactive contamination.

The scientists estimated that the explosion could occur on 8 or 9 May. For this reason they took every possible measure to extinguish the burning graphite in the reactor. The raging fire was eventually put out on 7 May.

Oksakovshchina has become the home to that special group of men called 'liquidators' or 'bio-robots'. These are the names popularly used to describe the 800,000 young men conscripted into the Chernobyl area from throughout the former Soviet Union to 'liquidate' or 'blot out' the released radiation. The intercession of these hundreds of thousands of young men – miners, soldiers, firemen – the 'liquidators' – is unparalleled in history. The self-sacrifice of these men cannot be overstated, as they prevented a highly likely nuclear explosion. To date, according to the governments of the three affected countries, at least 25,000 of these men have died. Many of the survivors became invalids. They were deemed heroes in 1986, but are now discarded and forgotten, vainly trying to establish that their ill health is the result of the extraordinary levels of radiation they were exposed to.

I visited Oksakovshchina in April 1993 to speak to these men, to hear from them exactly what their work at the reactor had entailed and to find out why they had been given these strange names.

The first man we spoke to had been in charge of the robots that had been brought in from Germany and Japan to remove the highly radioactive graphite from the reactor's core. The robots ceased to function because the levels of radiation were too high and interfered with their circuitry. That was when they conscripted the human 'bio-robots' to take the place of the now-defunct mechanical robots. These young men unwittingly exposed themselves to enormous doses of radiation. According to the survivors, the decision to use humans had not been based on scientific or technical grounds, but on political ones. The job was to be done at any cost and as quickly as possible. Many scientists objected strongly, but they weren't heeded.

On the night of the explosion, about 600 liquidators were at the site, mostly plant workers and firemen. They exposed themselves to horrific amounts of radiation. Twenty-eight of them died within three months due to acute radiation sickness. Many of them died agonising deaths miles from home in Moscow Hospital No. 6. A nurse talking to the wife of one of the dying fire-fighters said, 'You are young, what are you doing? What you have is not your husband, the man you love, but a reactor. Forget that that is your husband. That is a radioactive object – do not talk to that, do not touch, do not kiss or you will both burn to death.' Just by sitting by her husband's bedside she had become dangerously contaminated.

I have spoken with many of the parents of the fire-fighters who died and are buried in Mitino Cemetery in Moscow, and the one common plea from all of them is that their sons' bodies be returned to them for burial in their home places. I spoke to the mother of Vasily Ignatenko, the youngest fire-fighter to die. She was distracted in her grief. 'My boy, Vasily – I cannot live without him. Beautiful boy! Good boy! His soul can never be at rest as long as he is there [in Moscow] and not here where he belongs with his own people. How can I talk to him? I cannot afford to go to Moscow every time I need to talk with my boy. We never even held his body, to give him one last kiss; they said it was too dangerous and we might become ill. Imagine! Too dangerous to kiss and hold our own son! No, you cannot imagine the pain of that.' Vasily, like the others at the Mitino

A monument to V. Pavik, leader of t firefighters

18

Cemetery, was buried in a lead-lined coffin. Over the graves is a metre and a half of concrete slabbing, which is also reinforced with lead.

Ivan Shavre, now living in Narovlya, was a fire-fighter who survived. He was in one of the first fire brigades that fought the fire from the roof of the blazing reactor. He recalled, 'After about 40, 50 minutes of fighting there were two more explosions. There was a big black cloud, followed by an intense blue light. Then a ball of fire covered the moon. I felt sick and fell unconscious. I woke up in a hospital in Moscow with 40 other fire-fighters. At first we joked about radiation. Then we heard that a comrade had begun to bleed from his nose and mouth and his body turned black and he died. That was the end of the laughter.'

Tens of thousands of coal miners were conscripted because of their expertise in working underground and tunnelling. Their task was to dig out a tunnel under the reactor and to install a cooling coil for cooling the concrete base of the reactor and to reinforce any cracks appearing in the slab. The miners had to work in appalling conditions, not only in extremely high temperatures, but also very high levels of radiation. Their job was to prevent the block from disintegrating.

The conscripts received insufficient protective clothing to carry out their work, and in some cases none at all. Sometimes the men had to lift radioactive graphite with their bare hands. They fought over 30 fires that had ignited as chunks of the reactor were scattered over the Chernobyl site. They had to dig holes under the convulsing reactor, shovel highly radioactive hot waste off the adjoining roof and strip contaminated land. Much of this work was done with the minimum of protection. The men told us how officials had ignored radiation levels during the

19

clean-up and had even deliberately obstructed efforts by some to monitor the doses that the workers were receiving.

The 600 men of the plant's fire service and operating crews were the most severely irradiated, receiving radiation 13,000 times higher than the European Union maximum dose of radiation to which people living near a nuclear power plant should be exposed. Their heroic efforts to put out the fire at the reactor were eventually abandoned. On 27 April, in an attempt to quench the reactor fire, 30 military helicopters flew over the burning reactor and dropped 2,400 tons of lead and 1,800 tons of sand to try to absorb the radiation and extinguish the fire. Unfortunately, these efforts made the situation worse, since heat built up under the dumped material, making the temperature in the reactor rise again and increasing the radiation emitted. It took until 6 May to bring the fire under control and stop the reactor from leaking further lethal doses of radiation.

Gregory, a young soldier, spoke to me at Oksakovshchina. 'When we got there we were assigned to the roof and told to shovel. Those of us working on the roofs were called storks. We were proud to do it. We took a big slug of vodka and went over and back, over and back. We had to win, we were told. What? We didn't know! A war against the atom! They said in the papers we were heroes. My friend had to place the mark of triumph by placing the red flag over reactor Number 4. They showed it on TV in all its glory, but days later another volunteer had to do it again – the radiation had eaten the flag! We had to replace another and then another.'

The liquidators were also given the task of clearing contaminated villages of people and livestock. They had to shoot family pets. Entire villages were bulldozed and buried along with their contents. Contaminated soil was removed and buried and asphalt roads were laid in and out of the villages to keep down the dust.

The men spoke not only about their personal fears for themselves and their families, but also about the lackadaisical attitude of the authorities regarding the millions of acres of heavily contaminated land and the hundreds of uncategorised radioactive waste disposal sites scattered throughout the zone. There are 77 authorised nuclear waste dumps in Belarus alone, and because the dumps were being filled hurriedly in the midst of a chaotic and panic-stricken crisis, many of the sites do not meet the required safety standards. Little is known about the quantity and radiochemical composition of the waste placed into these dumps. Clothing, toys and cars were tipped into the huge clay-lined trench dumps. The men told us that when the dumps were full, they had to line up and leave row upon row of contaminated trucks, fire engines and military vehicles. Twenty years later, these vehicles are still standing, waiting in line. The men did their best, but there were very few people with scientific expertise to advise them about the dumps.

These brave men unknowingly exposed themselves to horrific amounts of radiation and are now paying a terrible price. One liquidator, Vadim, explained that since conscription he has suffered changes in his blood, has low energy levels, a bad liver and pains in his stomach and in the bones of his feet. While he was working, radioactive fuel had stuck to his skin. He is now waiting to die.

'While we were not able to understand everything, we saw everything. When the robots burned out we ran around doing their jobs in white suits and rubber gloves. They thought there would be a further explosion and needed to draw off the heavy water underneath in case the reactor would sink and then explode. So the water needed to be drained out and they asked us to volunteer to dive in there to open a hatch. Can you imagine that? We were tempted with promises of apartments and cars – some dived in – but the promises never came to anything. We were proud to do it anyway. It wasn't for money – it was to save our country. Many of them are sick now.'

A quiet, shy man, perhaps in his late twenties, spoke of his experience:

I went there too. I did not have to go. I volunteered for my country. I saw only the best of men there. At first, none of us were indifferent; we were proud of our work. It was later that I saw the men with vacant stares, after they had seen too much. Clearing the villages of people was awful. Like what you see in movies of the Great Patriotic War, except this time the village clearances were not being done by Nazis; it was us. We hated ourselves for that. On evacuating the villages, we would go back some time later to clear out the houses. Sometimes in the streets you would find maddened cows, dripping with milk, bellowing in pain. Agh! It was something terrible. I remember, I saw a cat in a window. I thought it was an ornament! Then I saw it was alive. We had to kill all the dogs and cats because they would carry the radiation, so I killed the cat. That first killing was hard. But I got used to it by killing from a distance, making no eye contact. Then it was easier. Sometimes we didn't get them all and they would run after our buses for miles. God forgive us.

In that same place we found many notes nailed on the doors of houses asking us not to loot. Some said things like, 'Be careful. We'll be back. Don't kill our cat because we have rats and we need him to catch them or they will eat everything.' Others had written notes telling what day and time they had left just in case some relation or friend would come looking for them. Some houses had poems written about the families who lived there. Others had painted signs saying, 'Our house, we are sorry for leaving you cold and alone!'

I eventually came home, many months later. My wife was very frightened and insisted that I throw all my clothes down the rubbish chute. I did that. All except for my militia hat. It had a badge on the front and my son, I knew, would like it. He was proud of me and went around wearing this hat. Some nights he wore it in bed! One year after that time he fell ill. It was a brain tumour. That was it. I can't say any more.

Another man, Vadim, told us about clearing the villages:

I handled the liquidation of five villages. I dream about that almost every night. It was terrible. We would arrive at six a.m., call to all the houses, reassure the people they would be returning. We told them to leave everything behind, but still they cried, they wailed like at a funeral, as if they knew they would never return. The houses were their whole world. They offered us moonshine and food so they could take maybe take the family samovar, a TV or a motorcycle. Yes, it was a barter system – everything was negotiable. I know we shouldn't do it but who can say? In some places whole villages moved! You can find so much of the stuff in the flea markets in Minsk and Kiev, even Moscow. We sometimes bartered cattle. I heard later they were sold in Russia very cheap – you see, they had leukaemia.

The old people were so trusting, accepting, their lives always having been an uncomplicated relationship with nature. Now that nature was dying. All those houses were like works of art, decorated in the most beautiful alive, bold colours, having sheltered and given life to so many. Empty now. The shadow of madness was on all of us.

Igor Shik with his liquidator ID card

A man called Igor Shik, a former laundry worker, had been conscripted to the Bragin region. Igor stayed for the minimum conscription time of 180 days, from 1 June to 24 October 1986. Many of his friends stayed for one year, since they had no jobs to return to. Igor and his team helped evacuate families, removed radio-active topsoil and cut down radioactive forests. Around Chernobyl, entire ancient forests were felled. Pines and firs had turned red after the explosion and the concept of a 'Red Forest' was born. They left one fir tree standing as a special memorial to 200 Russian partisans who had been hanged there by the Germans during World War II. Igor told us:

22

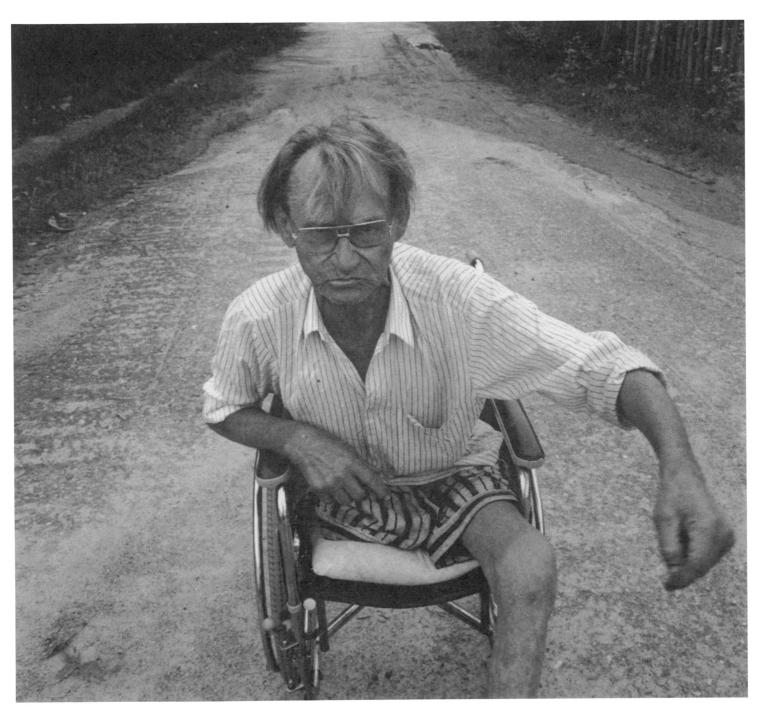

'You have to fight for your life every day, or you die' Nikadai Yanchin

We were told not to have children for five years because of our work. How do you explain that to your wife or girlfriend? Most of us didn't and hoped we'd be all right. We had to remove the top layers of soil and load it up on trucks. I thought the burial dumps would be complicated engineering places but they were just like open pits, not even lined with anything! We lifted out the topsoil in one big roll like a carpet with all the worms and bugs and spiders inside! But you can't skin the whole country; you can't take everything that lives in the earth. We stripped thousands of kilometres not just of earth but of orchards, houses, schools – everything. At night we drank so hard. Otherwise we couldn't do it. We slept in tents in beds of straw, taken from farms near the reactor!

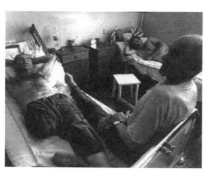

Igor also told us that once the job of decontamination had been done in a village, a special monitoring vehicle would drive through it to test the radiation levels. The men knew that their work had been successful if the two flags on the vehicle stood up straight. If the flags fell down, it meant that the levels were still too high. Follow-up checks would be made just days later, and the men would be dismayed to see that the flags that had been upright had dropped miserably, indicating that the area had been recontaminated. All their work, putting their lives at risk, had been for

nothing. In the meantime, the evacuated families had returned to what was still a radioactive village. When asked about the material used in the decontamination of the houses, Igor said that when they ran out of special chemicals, they used ordinary washing powder.

The liquidators were promised weekly check-ups. None took place. When their blood was finally tested, it registered a very high level of radiation. The testing ceased in order to prevent the men from panicking. At no stage were they given iodine (which would have protected their thyroid glands) in either liquid or tablet form, nor did they have access to a Geiger counter to measure the doses of radiation to which they were being exposed.

For months on end the liquidators slept in tents. Each night the tents were sprayed with some mysterious powder which was supposed to reduce the radiation levels. To prevent the radiation affecting their health, they were given extra food – one egg, milk, fruit juice and butter. As 'compensation', Igor and his friends were given an extra two weeks' holidays a year and free local transport if they lived in Minsk.

Igor has no home and currently lives in cramped conditions in his parents' flat. He has to live on the liquidator pension of €4 a month; there is little or no prospect that

he and his family will ever have their own home. His wife was seriously ill in hospital for over a year and he feels responsible for her illness. His son, born since the accident, is also extremely ill.

We questioned all the men as to why they had stayed in the radiation area once they knew it was dangerous. An older man, Victor, responded. 'Ha! It was like 1937! You know we have KGB here. Very active people! They tell us that we have families and they would be severely punished if we do not do our time. See, now? It's not so easy to escape. There was no escape. It was a game to them. They tossed us in there. It was a real war, atomic war. We had to sign a paper that we would not talk. After that time I became a Group II invalid.'

Another man told us:

You see, it was an exciting time for us. We were all gathered at the train station; it was in May, and we were told we had been chosen to do 'secret' work. We were honoured to serve our country, our motherland. The mood amongst all us conscripts was one of fun, like boys going on pioneer camp. There were some 7,000 of us reservists on the specially chartered train. We were feeling very important! Officially we were called 'tourists'. It seems like a sick joke now. From the trains we saw the fields change from green to something lunar – white dolomite sand covered miles and miles of field where the green earth had once been. We knew then something was very wrong. We lived in tents and the only thing that kept us going was the promise that once the Chernobyl block was cleaned up, we would be sent home. We were all happy when, after two weeks of demolishing villages, we were finally called to start work on the five roofs of the reactor.

Vitaly, a former liquidator, explains the experience of fighting fires at the Chernobyl plant:

The roofs had been given names by the men who worked before us, all women's names – Lyena, Marsha, Katya, Natasha and Anna. Oh, Marsha was the mad one. She was right over the exploded area and was cut open like a wound. We had to cross over several wooden gangplanks from one roof to the next and the faster you ran, the less radiation you got! All we had to wear was our cotton uniform, a lead apron, lead pants and a cotton helmet which had a glass visor.

We were living on the edge of our lives all the time. Imagine, we had only 60 and sometimes only five seconds to do our work on the roof! Any more time there, and we were exposing ourselves to lethal doses of the stuff.

You know something? I tell you this and swear it's the truth. We couldn't wear the lead pants because they cut into our legs and we couldn't run, so most of us didn't wear them. Anyway, we learned afterwards that the radiation was coming from down below in the belly of the reactor, up through our old boots, so the aprons made no difference. We stopped laughing very soon after we arrived in this hell place.

Many of us from the group started to have great difficulties breathing, so the running got

harder. We began to be dizzy and to vomit, and the balancing on the planks became harder and harder. Some of the men were taken away and we heard rumours they were in Moscow for treatment. I don't know the truth. We never saw those men again.

It's estimated that to date, 25,000 liquidators have died, and a further 70,000 are permanently disabled. Twenty per cent of these deaths were suicides. There has been inadequate monitoring and health checks for the liquidators, many of whom received over 10 times the recommended maximum lifetime dose of radiation in a few minutes. Geiger counters given to liquidators were confiscated by the KGB in order to keep the readings secret. All health information was officially classified by the USSR Academy of Medical Sciences. In July 1987, the authorities gave an order that acute and chronic diseases of liquidators who had been exposed to less than 50 rem must not be attributed to the effects of radiation.

A doctor who worked with the liquidators told us:

I was conscripted from 20 March until 25 September 1998. I worked with 13 doctors and 36 women nurses in the Bragin region. Our task was to check some of the liquidators who had been working the zone for prolonged periods. The situation was dreadful. We had little or no equipment or medicine and no Geiger counters. We stayed indoors for six months. We had no idea what the levels of radiation were. We knew it was high from the symptoms we found in the men and from the basic blood samples we were able to take. The men were from as far away as Siberia. None of them were allowed to contact their families, since phone calls were monitored and we were all warned about revealing where we were and what we were doing. It was a time in my life that I want to forget – the pain haunts me in my dreams even now.

Those of us living on the landmass of Europe need to be eternally grateful to the bravery and courage of the hundreds of thousands of liquidators who risked their own lives to save Europe from what would have been an extremely serious nuclear catastrophe.

The years are passing, but the memory of those who gave their lives and health for others will be with us forever. Lest we forget their heroism, I call on all people to remind ourselves of how we escaped this nightmare by placing flowers for them in a symbolic place on 26 April each year as our homage to the liquidators.

Vanya Chernousenko, a farmer in the Bragin region, gave this testimony:

How can you believe what you don't understand? On that day we did everything as usual. My son Igor and I milked our cattle, cleared out the cowshed. We looked after our cows well; they were like a family to us. At noon we went down to the place where we bring our cows every day for fresh water. They always love it, they know when we're going there and they used to run to it. But it was strange that day – the cows were subdued somehow, uneasy maybe. They did not run down this time, they seemed unsure or something. They bent down to the water but did not drink; they turned away without touching the water. They ran away. Even when Igor hunted them back, it was the same thing – they refused to drink. No one can explain what it was. But some days later we found out about the accident and we knew it was the reason. The animals were able to tell that something was wrong.

Andrei Smarlovski, a Khoiniki farmer, told us:

You see, they knew, the bees I mean, they knew something was wrong, but we didn't. Not until it was too late. I remember that morning well. I went out into my kitchen garden as usual; it was a lovely spring day and so beautiful. It was in full bloom; it's my very favourite time. The garden all dressed in wedding white. But something was wrong, something was missing, some old memorable sounds. Ah, you see, I realised something unusual, it came to me that I couldn't hear the

sounds of bees. This was something strange. My hives were over there, see, rows of them under the apple trees; they're rotten now, but not then – we had very good honey to sell and eat. 'What is it? What's wrong?' I said it then to Nina, my wife, and she said, 'It's a bad omen, Andrei; it's not right.' I put on my mask as usual and started checking the hives. They were there all right, sitting in the hives, not making a sound. There was no buzzing. So strange their silence, I thought. They are sick; maybe they have been poisoned from the fields, I thought. It was only when they came to take us away, three weeks later, that they told us there'd been an accident at the atomic station over there. You see, that is just 15 km from our kitchen garden. It's no distance for that radiation to come over here. We didn't know that, but our bees did. Something horrible happened so quietly and naturally.

Radiation knows no territorial boundaries. It doesn't apply for an entry or an exit visa; it travels wherever the winds take it. On 26 April 1986, a silent war was declared against the innocent people of Belarus, western Russia and northern Ukraine, a war in which they couldn't see the enemy, a war to which they could send no standing army, a war in which there was no weapon, no antidote, no safe haven, no emergency exit. Why? Because the enemy was invisible. The enemy was radiation.

The dense clouds of radioactivity were widely dispersed over Europe, with caesium and iodine directly contaminating

30

the lichen eaten by Lapland's reindeer, the grass of Wales and parts of Ireland and the flora and fauna of the French and Italian Alps. But the indirect fallout travelled much farther. In spite of the control measures eventually applied, it was soon discovered that in many parts of the Third World, dangerously irradiated food was being exported from Europe. There was outrage in July 1986 when Bangladesh discovered that a shipment of powdered milk exported from Poland and used to feed infants had over 100 times the permitted level of radioactivity. To make matters worse, infants are 10 times more vulnerable to the effects of radioactivity than adults. Over the following months, similar shipments from a variety of European countries carrying different produce were found to be radioactive. It became clear that tainted produce from countries in Europe affected by the fallout were being dumped in vulnerable parts of the world where controls and safety checks were inadequate.

On the warm, balmy days of early May 1986, 70 per cent of the fallout rained down upon Belarus. In the days following the accident, 99 per cent of the land of Belarus was contaminated to varying degrees by the radioactive fallout. This had devastating consequences. The land was covered by thousands of tons of caesium, iodine, lead, zirconium, cadmium, beryllium, boron and an unknown quantity of plutonium – in all,

40 types of different radionuclides. The fallout primarily affected rural areas – forests and wetlands areas as well as pastures and arable land. As a result, 23 per cent of Belarus's territory – 43,000 square kilometres – with 3,668 populated towns and villages were contaminated, mainly by caesium-137. In the Ukraine, 53,500 square kilometres, including 1,599 towns and villages with a population of 2.29 million people, were contaminated, and in Russia, 59,300 square kilometres populated by 1.78 million people in 1,088 towns and villages were affected. In Belarus, 2 million people, of whom 500,000 are children, still live in heavily contaminated zones of between 1 and 40 curies per square kilometre. These people are at the core of the cluster of problems created by Chernobyl.

Radioactive polluted lands are divided into four areas known as the 'contaminated zones'. The divisions are based on the work of Colonel Yaroshuk, a chemist and dose reader, who started at the reactor with a hand-held Geiger counter and worked out along a radius, marking off sectors according to the levels of radioactivity. The four zones cover an area of about 100,000 square kilometres. According to the Russian nuclear physicist Gregori Medvedev, 3 million hectares of land are unfit for farming, 2 million hectares should be kept out of agricultural use for 20 years and 1 million hectares kept out of use for 100 years.

On 27 April 1986, the first 'exclusion zone' was declared – a radius of 10 km from the reactor site. However, by 2 May it had to be increased to 30 km, as the authorities discovered that heavy contaminants had been carried to a wider area than anticipated. It became clear that, in fact, heavy contamination had been carried well beyond the 30-km zone.

Not only has radiation endangered the genetic future of those exposed to it, its power of destruction has also damaged the intricate cycle of all life. The suffering is medical, environmental, emotional, psychological and economical. The greatest resources of every country are its land and waters, and great damage has been inflicted on the natural resources, water management systems and agriculture of these regions, the most threatening long-term contaminant being the more volatile caesium. Where it is deposited in the soil, it will persist for years, particularly in the 5 cm of soil below the surface where plants have their roots. On forest floors there is severe contamination owing to the filtering effect of radiation stored in leaves, needles and roots. When the dead leaves and needles fall to the ground, radiation accumulates and penetrates down through the soil. In the coming decades it will also accumulate in the wood of the trees. The speed at which this occurs depends on the type of soil. In clay and sandy soils, the downward migration of caesium is very slow, but in deeper layers of peaty soils, the process happens much more quickly. The effects of this are found in typical forest plants such as heather, lichens, mushrooms and ferns. Other grasses and plants are also affected to varying degrees, depending on the type of root and soil. For example, if the plant has a shallow root, it's more severely affected than if it takes its nutrients from deeper soil layers. Where soil is poor in minerals such as potassium, plants will quickly absorb the caesium present since chemically it has a similar nature to potassium.

The land formerly known as the 'breadbasket of Russia' is now poisoned with contamination from the fallout. Some 480,000 hectares of farming land, including 230,000 hectares of arable land, have been withdrawn from agricultural production. The territory adjacent to the river Pripyat, once a major meat- and dairy-producing area, has been turned into a depopulated radiation zone. The river is

trance to the exclusion zone

hazard to the groundwater. However, the main problem relates to internal irradiation resulting from the consumption of contaminated foodstuffs, namely milk, meat and forest food such as mushrooms, fish, berries and game. This threat can be partially controlled by the use of special fertilizers and fodder supplements, but because of rising poverty, few people can afford the necessary precautions.

One of the most severe problems is soil clean up and decontamination. While there was a large removal of topsoil over a 1,000 square mile area both inside and outside the exclusion zone, it is still not known, 25 years after the clamity, what was done with it. Many doctors are not so concerned about the gamma radiation that passes through the human body (although that kind of radiation is still too high in some areas) as they are about the lethal radioactive particles of caesium, plutonium and stronium in the topsoil and roadside dust, ready to be stirred up, inhaled, ingested or eaten in contaminated food.

The radioactive contamination of the land isn't the only danger posed. The use of chemicals in the attempt to reduce radiation has had an effect on the natural ecological balance. There isn't enough uncontaminated farmland left to feed the population and, since caesium-137 continues to stay in the upper ground level, it will be accessible to plants for a long time. Strontium-90 half-transformed

highly radioactive, and since it feeds into the river Dnieper in the Ukraine, it has caused the silt bed of the Dnieper to become radioactive. The Dnieper is now the world's most radioactive river.

In the 30-km area, large quantities of soil were removed to 600 to 800 unsecured burial pits. These radioactive dumps were never precisely mapped and pose a huge

in a free form becomes easily accessible to plants and enters the food chain, eventually getting into the human body and increasing the risk of health problems.

Decontamination isn't producing the hoped for results, since the radioactivity is spreading at a frightening rate throughout the country and beyond. Radioactive substances from the Chernobyl accident are spreading out into the environment much more quickly than was expected. According to the Russian geological chemist Waleri Kopejkin, even the Ukrainian capital, Kiev, is under threat: 'If the Ukraine would implement the international accepted standards for strontium-90, we would have to evacuate Kiev.'

Evgeny Konoplya, director of the Radiobiology Institute at the Belarusian Academy of Sciences, says that the land cannot be cleaned: 'You would have to remove the entire fertile upper layer of soil, tear out trees by their roots and turn these areas into a desert.'

It has become impossible for farmers to know if their land and its produce are 'clean' or not. Some areas will be radioactive for 24,000 years, which is the half-life of the highly radioactive material plutonium.

Because of the severe economic situation in Belarus, the government is unable to supply its population with imported 'clean' food, and so the people have no choice but to eat the food produced by their own farmers. The main source of contamination is the food chain. Between 20 and 30 per cent of milk consumed by the population in the zones contains unacceptable levels of caesium-137.

Five of the six regions of Belarus are deemed to be contaminated, but the farmers are forced to continue agricultural production. Much of the food produced turns up in marketplaces all over Belarus, including the capital, Minsk, where local people and tourists buy it unwittingly. Within the contaminated zones there are 3,678 inhabited towns and villages, most of which will never be evacuated.

Radioactive particles from Chernobyl were deposited on the soil, water, vegetation, buildings and on entire communities. These particles were the major component of 'external' doses of radiation received by the population. At the time of the initial fire, the weather was showery, so the distribution of the fallout was largely dependent on where it happened to rain. This explains what later became known as 'spotty' or 'patchy' contamination. It also explains how the levels of contamination can vary between one village or town and another, and even between adjacent fields.

According to many experts, there continues to be major problems for those living in the zones due to the constant danger of secondary contamination of the soil by airborne radionuclides which occur

in dust storms, bush/brush forest fires and on peat beds. In addition, substantial amounts of radioactive material locked in soil or vegetation are released into the atmosphere and transported across large tracts of land, adding to further radioactive movement. With the new government policy of refarming in highly contaminated areas, this problem is exacerbated by radioactive dust generated from ploughing. We witnessed this when filming in the 30-km exclusion zone in 2004. In effect, recontamination is taking place with further amounts of such elements as strontium-90 and caesium-137. The most dangerous time of year for such occurrences is during the dry, hot summers.

The people living in these areas are constantly exposed to increased internal doses of radiation through inhalation from forest fires, as well as from the radioactivity that comes from the burning of contaminated wood for cooking and heating.

I remember being on a convoy in April 1996 when Ali Hewson (our patron) and I, along with Irish journalist Frank Shouldice, were in the Children's Regional Hospital in Gomel, which is in the heart of the Chernobyl region of Belarus. It was about eight p.m. on a warm evening and all the windows of the hospital were wide open. A call came through for Ali on the mobile from a radio station in Ireland, saying that there was a dangerous series of forest fires raging out of control in the

exclusion zone, about 80 km from Gomel, and due to the wind direction, it would be over the city where we were delivering aid within a matter of hours. Were we considering pulling out our aid workers? The movement of radioactive particles from the land to the air meant that we would be exposed to the risk of inhaling or ingesting radioactivity carried by the winds moving swiftly in our direction. The medical staff made several phone calls and sent a doctor to the local chief of police, but nobody knew anything. We then had the problem of tracking down six of our trucks and 15 ambulances, which were carrying aid workers delivering the aid. We managed to locate all vehicles and workers, who arrived back to the hospital to work out the next stage of our response. We all gathered in the grounds of the children's hospital. The workers were jumpy and nervous, acutely aware of the fast-approaching danger being carried by the winds. Two hours had passed since we got the news. From then on, every extra minute lost meant we were placing ourselves closer to the risk of exposure.

We decided to travel in convoy, instructing the drivers to keep windows tightly shut, to switch off air vents and not to stop under any circumstances until we were back in Minsk, a five-hour drive. Many of our Belarusian associates had gathered with us in the yard. They were visibly frightened and upset. I felt guilty for having to abandon them as we gave each other final warm embraces. Valentina Pokhomova, the local leader of the Chernobyl organisation Children in Trouble, said, 'This is how it is, our lives always in the balance, no one will know about tonight here. Only because you are foreigners you have been made aware. Look, all the windows of the children's wards are open wide, meant to keep the children nice and cool during the night. But you know, maybe in three hours it will bring the sleeping, innocent children the air of death. What to do? We can do nothing. Go, you leave here, be safe and tell the world what is happening. Tell them how you too almost became one of us, victims of Chernobyl.'

The final image of that night is frozen in my mind – a small group of people standing stoically and lost, unable to prevent the inevitable, with the backdrop of three floors of open windows, bringing a gentle cool breeze, lulling the children to sleep. I didn't know that death could be so beautiful.

In addition to rainfall, the rivers carried the radiation on their surfaces for days after the accident, so that there was the added dilemma posed by contamination via river networks, lakes, ponds and dams. This is particularly a problem in the case of strontium because it's more mobile than caesium and readily soluble in water, making it more difficult to track. The main fallout was concentrated in the watershed territory of the rivers Dnieper, Pripyat and their tributaries.

There was no means of decontaminating the marshes and swamps in the common border region between Belarus and Ukraine. In Belarus this area is called Polesye (the Woodlands) and is known in English simply as the Pripyat Marshes. This forested wetland of approximately 100,000 square kilometres will remain contaminated for several decades with substantial doses of radioactive caesium and strontium, which are quickly absorbed into plant tissue. Many of those who are too poor to move away wade chest-high in the bogs to harvest cranberries.

In the wintertime, the snow has the effect of suppressing the radioactivity and locking it underneath the ice and snow. However, problems arise when the big spring thaw comes. The resulting radioactive water penetrates the soil, rivers and streams, carrying the radioactivity through everything in its path. Later, in the intense heat of the summers, the soil becomes dry, and light winds carry the radioactivity throughout vast territories of farmland, rivers, towns and villages.

Floodplain lands also come into the equation as another source of contamination of surface water. The runoff and washout is a more serious long-term problem from lands upwind of the reactor, which feed into the reservoir of the Dnieper system. The 30-km exclusion zone encompasses a number of aquatic

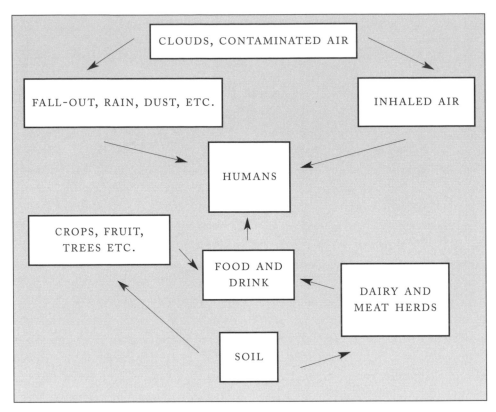

ecosystems already subjected to considerable contamination from the regular
operation of the Chernobyl plant, including the northern part of the 1,000
square kilometre Kiev reservoir behind the
Kiev-Dnieper dam, as well as a section of
the Pripyat River. This poses a huge threat
to the drinking water of millions of people
using downstream water in the Ukraine
and all the way down to the Black Sea.
Added to this is the constant fluctuation
of elements like strontium-90, which are
highly mobile.

The pathway via the land and water
systems to the human body is external via
the food from the soil, from swimming and
fishing (particularly with bottom-feeding
fish, which are heavily contaminated) in
the lakes, rivers and ponds, from the food
of both the water and the land and through
drinking the water.

The land and water crisis is further
exacerbated by the patchy distribution of
the radioactive fallout and by the ways in
which it was deposited. Rainwater often
swept it into the ground, and therefore it
has persisted in the soil and consequently
in the food chain. The authorities have set
a 'lifetime dose' of 'permissible' radiation
for those left living on the land. This was

initially set at 25 rem, but the figure was altered in 1988 to 35 rem. In contrast, according to the International Commission on Radiological Protection, the lifetime dose limit set for members of the public is 7 rem. Strategies need to be developed for tackling the dispersal of radionuclides in water and into the atmosphere. In addition, the implications that this has for families, hunters and forestry workers, who are all at risk because of their particular lifestyle patterns, need to be addressed.

Despite the natural decay process of radioactivity, the pollutants from Chernobyl will pose a health hazard for decades. Some of the long-lived isotopes, particularly plutonium-239 and americium-241, will stay in the environment for thousands of years. Some scientists say that it will be a further 50 years before we see the full impact of Chernobyl both medically and genetically, but by then it will be too late for countless people. The short-term ambitions and political expediency of the few will blight the future of generations.

In 2002 we started to see a shift in political thinking about the impact of the accident. The new approach appears to hinge on 'blotting out' the radiation and remarketing the impact. This shift started with political comments that the problem was now resolved and that somehow, miraculously, the radiation and its effects were no longer a consideration. It began with marketing the country as a tourist resort, with an emphasis on endless hunting possibilities within areas that, until now, had been deemed 'contaminated'. At around the same time, farmers were being actively encouraged to return and refarm their contaminated land. The state now offers a house and a job in a collective farm or a dairy plant. Today you can find in previously abandoned villages the new 'pioneers' who are taking over abandoned homes.

One evacuee farmer, Vasily, originally from the Khoiniki region in the exclusion zone, spoke to us: 'I don't know what to do any more. I'm tired of being told this and that. The officials who told us that we would never return to our land are now giving us orders to go back and start ploughing those same fields or else we will be put in prison. Some of us have gone back because we are afraid of being sent to prison. They are checking, they told us, by helicopter and will know which farmer is working his land and which one is not. So we cannot escape. What to do now?'

Seventeen years after the accident, a resolution was passed by the Council of Ministers of the Republic of Belarus that redefined the status of many of the villages in the contaminated zones. This political decision has had an appalling effect because 146 villages, home to 66,000 people, including 17,000 children, are now excluded from the special 'Chernobyl zone' status. The impact for those living in these villages is that they are now deprived of the special aid that is normally granted to Chernobyl zone areas. It also deprives the children of free nourishment at schools and kindergartens, and they are now exempted from their regular visits to sanatoria for medical care and rest. According to the resolution, the radiation status was reduced in a further 71 villages with 60,000 inhabitants, including 13,000 children.

Local government authorities asked the Belarus Institute of Radiation Security (Belrad) in Minsk to measure the levels of caesium-137 in the inhabitants and their food in a number of settlements in an effort to determine the true levels of contamination and therefore the degree of danger from living in such areas. They were seeking to provide the necessary radiation protection for the inhabitants. These measurements were carried out in 20 of the newly exempted villages (villages with 'no' Chernobyl effects) and 17 villages where the radiation status had been lowered. What these scientific measurements categorically showed was that, in each case, caesium was detected.

They give conclusive evidence that all the settlements should have remained within their previous status. The state had understated the levels of radiation by six to eight times. What is so worrying about this new trend is the attempt to scale down the situation with the long-term objective that it 'disappears' altogether. By declaring countless towns and villages to be 'clean', the authorities shift responsibility, thus abandoning people to their own fate. The future of the 146 villages, now known as 'the forgotten villages', home to 74,290 people, including 24,000 children, remains in the balance. Independent monitoring by Professor Vasily Nesterenko clearly shows that the basis for the resolution being passed was both flawed and inaccurate. The result for the inhabitants is the reduction or abolition of any protective measures, such as nutrition for children with clean food products in schools and kindergarten and the reduction of rehabilitation programmes for adults and children in the agro-industrial complex. It now appears that there is a concerted effort to blot out the consequences of Chernobyl.

Professor Nesterenko

42

RADIATION EFFECTS ON HEALTH: A GENETIC TIME BOMB

ow-level radiation creates the
ndition in everyone for many cancers,
eases and defects. It's getting worse.
e have to find a way to live safely in
se dangerous conditions. Our people
ed protection from their world. We
ve to find ways to protect our patients
d the people in the future.'

*Tamara Belloknya, director of the
sakovshina Sanatorium*

Radiation reached the people of the three affected countries externally when they were exposed to radioactivity in the initial cloud or through radioactive elements deposited in the soil. The internal exposure occurred through inhalation of the dust from the cloud or ingestion via contaminated food. Internal radiation causes the most damage.

The element caesium-137 is the most important radionuclide to consider because it's responsible for most of the radiation exposure received by the people in Belarus, western Russia and northern Ukraine. Before the Chernobyl accident, deposits of caesium-137 on the land mass of Europe was primarily the result of global fallout from the atmospheric testing of nuclear weapons.

After extensive studies, Professor Yuri I. Bandazhevsky, the former rector of the Medical Institute in Gomel, stated that between 1990 and 1999 the biggest health threat was caused by the incorporation of caesium-137 in the human body. Once integrated in the body, the radioisotopes influence and can change the metabolism of tissues and vital organs. He added that there should be no caesium in the body and there should be no question of temporary or acceptable levels. Levels of caesium in the human body were inadmissible, he said.

Professor Bandazhevsky and his wife Galina, both physicians, moved to Gomel after the accident because they felt it was their duty to offer expertise to those living in the region. During the 1990s they

43

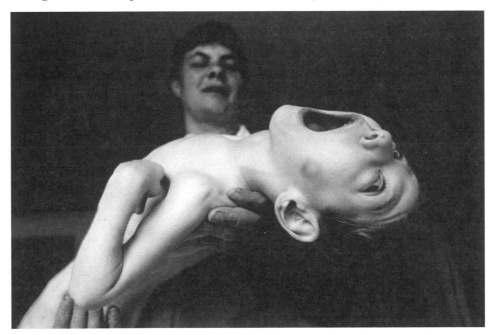

noticed an alarming increase in heart problems and birth defects among children in Belarus. The couple saw children as young as six years of age suffering from strokes and heart attacks. They began to keep statistics of their patients, and with government funding they and others at the research institute in Gomel where they worked began to research the impact of radiation on the human body. They designed and produced instruments that could measure caesium in the body and in food. Based on their extensive research and analysis on the amount of caesium that children were ingesting, they started to advocate the necessity of clean food products for children.

Professor Bandazhevsky shows the correlation between the easy penetration of caesium into the soil and how it becomes absorbed by plants and then incorporated into the animal and human body through food and water. Human absorption occurs when fruit, grains, meat, vegetables, milk, fish, mushrooms, berries or water are ingested. The intake of caesium varies and depends on the sex, age and physical condition of the animal or human at the time of absorption. Cells of different organs intensively incorporate caesium, primarily the thyroid and other endocrine organs, as well as the heart, kidneys, liver, brain, spleen, bones and muscles.

Professor Bandazhevsky has also discovered that incorporation of even a

44

small amount of caesium has the effect of modifying the metabolism of both the kidney and heart organs. Through his specific heart research he discovered a life-threatening heart disorder called caesium cardiomyopathy in people from the contaminated regions. It also enhances the possibility of innate diseases by in-fluencing the hormonal relationship in the mother-placenta-foetus system. Women of childbearing age who live in the contaminated areas consume strontium-90 via food. It can pass through the placenta and into the foetus, where it can lodge in the bones and bone marrow. The developing cells in the embryo/foetus are particularly at risk and the mother's exposure to radiation can lead to severe organ and brain damage in newborn babies. Thus we find the answer to so many previously perplexing problems, particularly in newly born and young babies, including significant increases in the incidence of stillbirths.

Professor Bandazhevsky further discovered that large amounts of caesium also influence the blood and immune system. In this case, the breach level is in direct proportion with the amount incorporated. In his writings he states, 'It is much easier to prevent the radio caesium penetration in the human organism than to excrete it and to correct the possible illnesses caused.' He continues by remarking that the biggest damage is done to children: 'The health

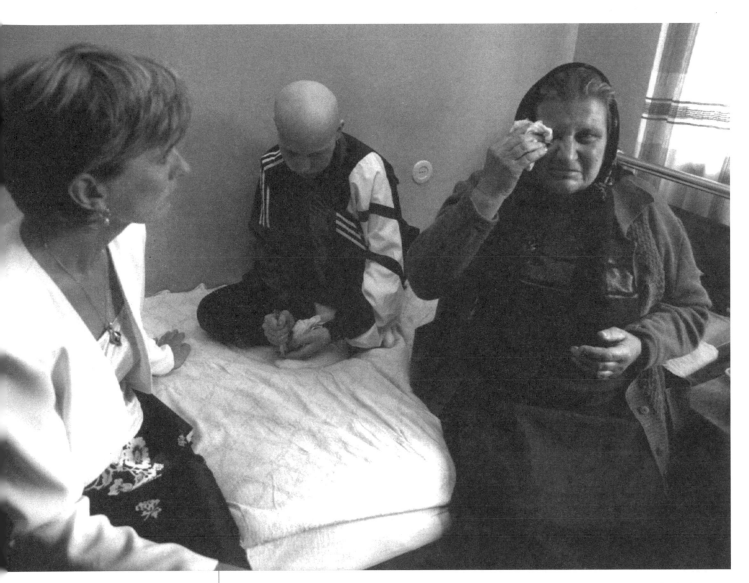

condition of the affected population is a disaster, but being a physician myself, I cannot accept the situation as hopeless. With all my faith in God and life, I appeal to anyone who can help: do your best to improve the situation. There is nothing more precious in this life than life and we should do everything we can to protect it.'

Professor Bandazhevsky has suffered greatly for publishing his research. He was arrested on 13 July 1999 and sentenced to eight years' hard labour and imprisoned on

18 June 2001. He was later adopted as a prisoner of conscience by Amnesty International. Professor Bandazhevsky has committed no crime. The only thing he is guilty of is telling the truth. He has always stated that he wanted to teach people 'how to live under the shadow of Chernobyl'. Unfortunately, the state of Belarus wants to pretend that the shadow doesn't exist. Professor Bandazhevsky was eventually released from prison in 2004, having served three years, but was then placed in a special compound-type village set up for prisoners, where he remained under strict supervision. He had little or no contact with the outside world until July 2005, when he was released. However, Professor Bandazhevsky's freedom is still restricted and he isn't permitted to travel abroad.

I cannot keep count of the children I have met and who subsequently died – died, I believe, as a direct result of Chernobyl. Children share the same menu as adults but are most at risk since they receive five to six times more radioactivity than adults owing to their smaller weight, height and more active metabolism. In general, it's the rural community rather than those living in towns and cities, both adults and children, who are hardest hit because they are almost completely dependent on local foodstuffs for their survival. The survivors are facing a demographic disaster where science cannot yet completely assess the consequences. Dr Vladislav Ostapenko,

head of Belarus's Radiation Medicine Institute, reports that 'we are now seeing genetic changes, especially among those who were less than six years of age when the accident happened. These people are now starting to have families, so we are witnessing the effects of the disaster move to the next generation.' The silent killer, radiation, is threatening the gene pool and the future of the people.

Countless lives in the Chernobyl-affected areas have been destroyed by ill health. What we are witnessing is the slow erosion of a nation's health. Scientific research shows that the people are faced with soaring levels of infertility and genetic changes, affecting the very future of their race. Their mortality rates already outstrip their birth rates. According to the UN, 7 million people are affected, half of whom are children. In Belarus alone, 90 per cent of children are deemed to be victims of Chernobyl. A general increase in morbidity from non-oncology conditions such as cardiovascular and respiratory conditions are adding to the general decline in health of the people living in the affected countries.

It's difficult to obtain a true assessment of the impact of the tragedy on the health in the three affected countries; data from experts and health professionals from the region isn't always believed by their counterparts in the West. It hasn't been in the interests of some that the true health toll be known. In Belarus, the computer files holding the health data on all accident victims simply disappeared.

Because of severe damage to the immune system, there is deep concern for the body's inability to fight cancer cells. I remember when this became real for me while visiting an oncology hospital in Minsk. I was with a team of doctors in the children's ward and came across a very ill little girl of about two years of age. Her skin was like parchment, grey in colour, sticky and clammy to the touch. Tamara was as light as a feather and seemed to be slipping in and out of consciousness. Her mother sat beside her, continuously stroking her hand. I asked the doctor what the diagnosis was and he replied, 'There is absolutely no hope here. This child is dying. Yes, dying. And why? Well, it is because she (sideward glances at the mother) breastfed her as a baby. Can you imagine? What stupidity! To breastfeed when that milk is destroyed with radiation. So stupid.' I could have hit him for such dire insensitivity. The impact of his brutal words was swift. The mother of the child, stricken with grief and guilt, collapsed over the bed, holding her little

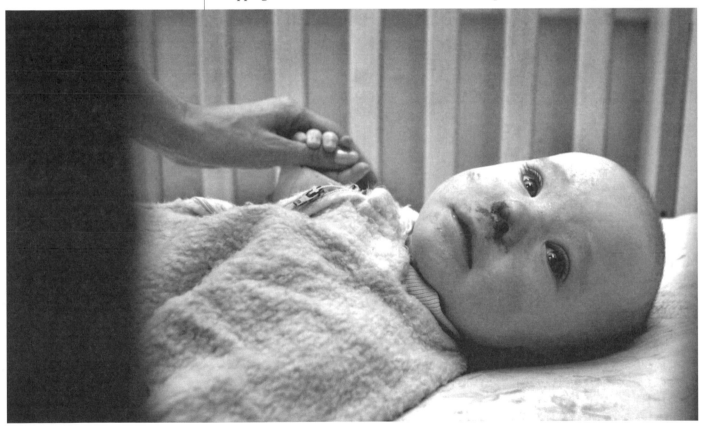

child. Tamara died later that night. According to Professor Nesterenko, only 7 per cent of mother's milk is not radioactive, and he considers it better for women in the region not to breastfeed their babies.

The most significant short-term contaminant isotopes are the iodine ones, in particular iodine-131, which has a half-life of eight days. It's either inhaled as a dust particle or ingested in cow's milk. Once it finds its way into the human body, it's mainly absorbed by the thyroid gland, which also intensely incorporates caesium-137. Since infants and children have more active thyroids than adults and also drink much more milk, they become the high-risk category group. Most of the patients now suffering from thyroid problems were in utero and beyond 11 weeks of gestation or under the age of six at the time of the accident.

I spoke to Dr Evgeni Demidchik, director of the Thyroid Tumour Centre in Minsk, who has become the world's leading expert on thyroid cancer because of the Chernobyl experience. His special concern is thyroid damage among children. For eight days after the accident, iodine-131 was breathed and consumed in food by the entire Belarusian population. In northern Belarus, 250 miles from Chernobyl, the concentration in the air surpassed admissible levels by 1,000 times. Those who have had a full thyroidectomy will require follow-up medication for the rest of their lives. While the medicine is relatively inexpensive to us in the West, it's a fortune for the people in the region. I have often been in the disturbing situation where people have come seeking us out in the hope that we might have a supply of L-thyroxin to give to their children, who have become dependent on it for life.

Already Belarus has seen such soaring levels of thyroid cancer that it's considered to be of epidemic proportions. Nearly a quarter of all children up to three who were exposed in the highest fallout areas can be expected to develop thyroid cancer. Prior to the accident, thyroid cancer in children was almost non-existent. According to Dr Demidchik, thyroid cancer has increased by 2,400 per cent. In the Gomel region there is a hundredfold increase. Dr Demindchik says, 'It is as if our children have been attacked by radioactive iodine 131.'

The incidence of thyroid cancer has also increased among adults. Initially this Belarusian thyroid cancer crisis was treated with scepticism until the World Health Organization (WHO) verified these findings in 1995. The increases in the three affected countries represent a unique situation in which a single cause, at a particular defined time, has resulted in huge increases in a specific cancer. Previous experience of irradiation of the thyroid indicates that cases related to radiation exposure will continue to arise for at least 50 years after that exposure.

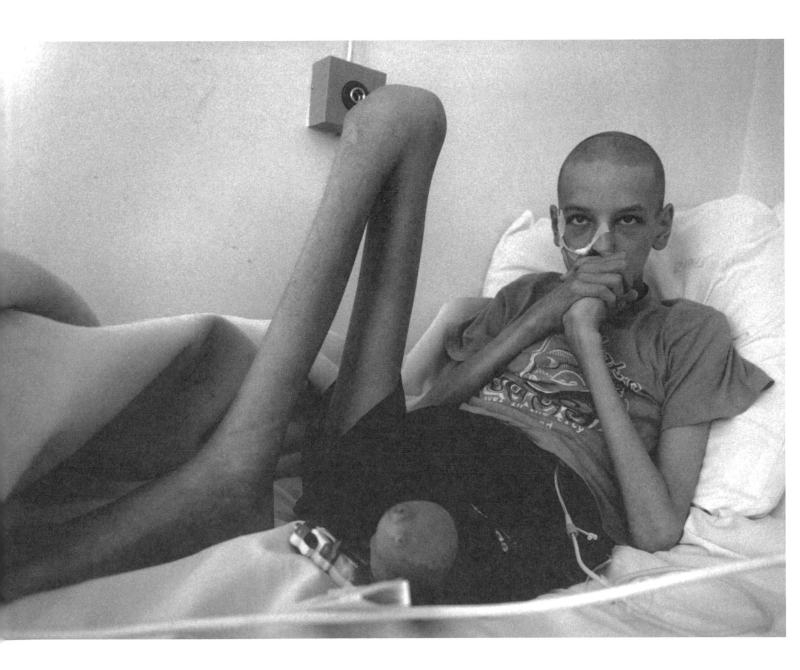

I was with Dr Demidchik and his team moments after they had told a young teenager, Ira, that they had detected eight nodules in her thyroid. She was now facing a fine needle biopsy followed by intense treatment in an attempt to save forsaken our family? It is my fault. Ira is this way because we were too poor to move away from that place. I cannot forgive this. I will never forget what Chernobyl is doing to us. My Ira is doomed. Chernobyl is Sveta, Sveta is

her life. Her mother, Sveta, was crying by her bedside, holding her daughter's hand. They were inconsolable. Etched on both their worn faces was a look of utter despair. I tried to offer some words of hope. I had known this family for many years and wanted to be of some support. Sveta reached out and said, 'Why has God Chernobyl. What can I do?' What could I say to that? I had no answers. All three of us enveloped each other with our arms, swaying, crying, hoping for a miracle.

Dr Demidchik has conducted the most comprehensive study of the development of thyroid tumours in Belarus, a study that began in 1966, when thyroid cancer was a rare disease. After the Chernobyl disaster, he began to see a marked increase in cancer of the thyroid and thyroid abnormalities in children. In the two decades before Chernobyl, there was less than one case per year. From the time of the disaster up to the year 2000, 7,504 children and adults had received surgery for thyroid cancer. All these new cases came from the most contaminated regions – 30 per cent came from the town of Pinsk in the eastern part of the Brest province, and over 60 per cent came from the affected parts of the Gomel region. All the children surveyed were born shortly before or during the time of the disaster. According to Dr Demidchik, because the native diet was already seriously lacking in natural iodine, the population's thyroid glands thirstily absorbed the radioactive iodine 'like sponges' because the body was unable to determine the difference between radioactive and natural iodine. Where potassium iodine tablets were promptly distributed to children – for example, in the Mogilev region, a heavily contaminated area – the impact was far less severe.

The tumour must be discovered and removed at an early stage. The Belarusian procedure involves the removal of only the affected area and not the entire gland, which is the common practice in the West. Dr Demidchik says that thyroid tumours caused by radiation are very aggressive and the cancer spreads rapidly to the lymph glands and to other parts of the body. Early diagnosis is imperative if lives are to be saved. The state-appointed Belarusian Chernobyl Committee predicts that there will be a further 15,000 new thyroid cancers in the next five decades.

The incidence of this disorder has considerably exceeded the world level and its rate of increase is surpassing all the predictions of the world's experts. According to Professor Edmund Lengfelder, Professor of Radiation Biology Medical Faculty in the University of Munich, 'Thirty per cent of all children who were exposed from the area around Gomel who were aged between zero and four at the time of the nuclear disaster will contract cancer of the thyroid during their lifetime. In this region alone, this is 50,000 people.' This information isn't well known, either locally or in the West. Perhaps one of the reasons for the lack of information is that iodine-131, the principal thyroid-seeking fallout isotope, is discharged by the nuclear industry in the West. It's also the abundant isotope found following nuclear weapons tests. In the 1950s, the US Department of Energy deliberately released a cloud of iodine-131 as an experiment to track the ensuing plume. The American government is constantly being petitioned for compensation by cancer survivors who live or have lived near the Nevada nuclear weapons test site or other nuclear facilities. Fears of crippling compensation claims, entrenched scientific opinion and concern about bad public relations for the industry have all added to the resistance in the West of accepting the findings of scientists in the affected countries.

Only when the WHO broke the international silence on damage to the thyroid gland by issuing the results of findings by the WHO and scientists from Belarus, Ukraine and Russia on increases in childhood thyroid cancer was credence finally given to the mammoth pioneering work of Dr Demidchik and his team. A clear link has now been established between the regions contaminated by radiation from the Chernobyl accident and the increase in childhood thyroid cancer. Importantly, thyroid cancers are indicators of other diseases, such as breast cancer, that are likely to follow.

The International Federation of the Red Cross and the Red Crescent operate six mobile diagnostic laboratories, which travel throughout the remote areas affected by radiation. Their sole task is to check for thyroid abnormalities. The local health authorities have little, if any, capacity to bridge the gap and detect, diagnose and treat thyroid problems, so the Red Cross Red Crescent programme is of life-saving importance. The annual screening programme has so far examined 600,000 people and aims to examine upward of 90,000 people a year.

Thyroid cancer is one of the early cancers that has damaged so much of the population. However, the majority of other cancers have a latency period lasting for up to 30 years and more, so studies and research must be continued into the future.

Behind all these statistics is a terrible suffering. We did some filming in the Thyroid Tumour Clinic in Minsk, which was filled with young children and adolescents. Many of them had the tell-tale bandage across their necks. When it would eventually be removed, they would have the permanent looped scar from ear to ear, the so-called 'Belarusian necklace', marking them forever as Chernobyl victims. In their cramped rooms, the children shyly told us how much they were missing their family and friends. What disturbed me was to see children with their glands recently removed and in obvious pain and distress, side by side with children awaiting the same fate, their proximity to each other making the

anxiety of the others even stronger. Child after child spoke to us about their friends at home who had suffered the same illness. There was a fatalism in them, particularly in the teenagers, because they had a greater awareness of their plight.

We visited many other hospitals and spoke to many doctors, who told us there had been a marked increase in other health effects. These included disorders of the endocrine system, major organs (heart and kidneys), the nervous system and sensory organs, the immune system and genes in cells, congenital heart and circulatory diseases, disorders of the bone, muscle and connective tissue system, and malignant tumours. The incidence of diabetes mellitus, where patients become insulin dependent, has increased dramatically. Blood disorders and disorders of the blood-producing organs have increased, along with problems with the nerves controlling circulation and problems related to eyesight. An increased incidence of breast cancer as a direct consequence of the accident has also been recognised internationally. The number of cases has doubled in the Gomel area of Belarus.

The results of the International Federation of Red Cross and Red Crescent medical screenings in the contaminated regions are disconcerting. In six regions of Belarus, Ukraine and Russia, the Red Cross mobile diagnostic laboratories screened 28,585 people between October and December 2000. Of the

53

10,946 adults screened, 9,091 (83.1 per cent) were found to be sick; of the 17,639 children screened, 13,533 (76.8 per cent) were found to be sick.

These figures should make us think, particularly when much of the damage to people's health could have been prevented if proper treatment had been taken. When I interviewed leading Belarusian nuclear physicist Professor Vasily Nesterenko, he told me, 'The authorities will have to answer for Chernobyl one day. Explain how they had 700 kilograms of iodine in warehouses, and that is where it stayed. Except, of course, they took it for themselves and their families! I know this because when we checked their thyroids, they were clean! Those in charge were more afraid of Moscow than they were of the power of the atom! They all waited for a phone call, an order, but none would take the responsibility himself to do something. They had special clothing and respirators for themselves too. They even had a special herd of cows outside Minsk for their own personal milk. The cows each had a number, which corresponded with someone's name. They had their own special greenhouses for food. It is disgusting. And yet no one was brought to account for these despicable acts. I deal only with facts. I have plenty of facts. I set up my own monitoring. Used the best equipment to find out the truth. I have made maps tracing the radiation path. My colleagues and I tested the first children in the villages. They measured 1,500, 2,000, 3,000 micro roentgens. Those girls will not be able to have children. They have genetic markers now that we cannot do anything about. It is a genetic time bomb.'

There is huge concern over the related health problems resulting from immune system damage. Collectively, doctors often refer to 'Chernobyl AIDS' as the overall name of the resultant depletion of the immune system that is caused by the effects of strontium and accepted by the body because it has a close resemblance to calcium. The body cannot detect the deception created, so strontium gets absorbed by the bone structure of the body and remains there for a long time. Strontium is highly radioactive and has a disastrous effect on bone marrow and bone growth in children during their growing period. It takes anything up to three years before the damage caused by strontium becomes visible. The weakening of the immune system that strontium causes can probably explain the rising number of cancer cases. Many doctors have expressed a deep concern about the current increases in different cancers and about the cancers which are already present but not yet evident. These cancers are now able to break through because the body's immunity against cancer cells has been seriously undermined and damaged.

The health deterioration in Belarus in the post-disaster period is characterised by a persistent growth in the morbidity rate

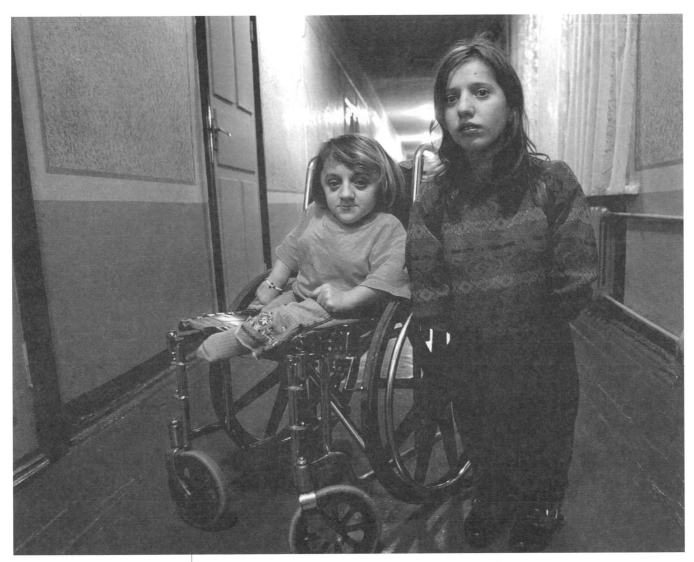

among children, especially of younger ages, as well as amongst pregnant women and women of childbearing age. It's also characterised by an increase in birth disorders and a growth in infant mortality. In the radiation zones, the birth rates have dropped by 50 per cent. For the first time in Belarusian history, the death rate is higher than the birth rate. All the women I have met and spoken with constantly reiterated their desire to have children – healthy children – but their fears of illness and damage were too great and overwhelmed their urge to have children.

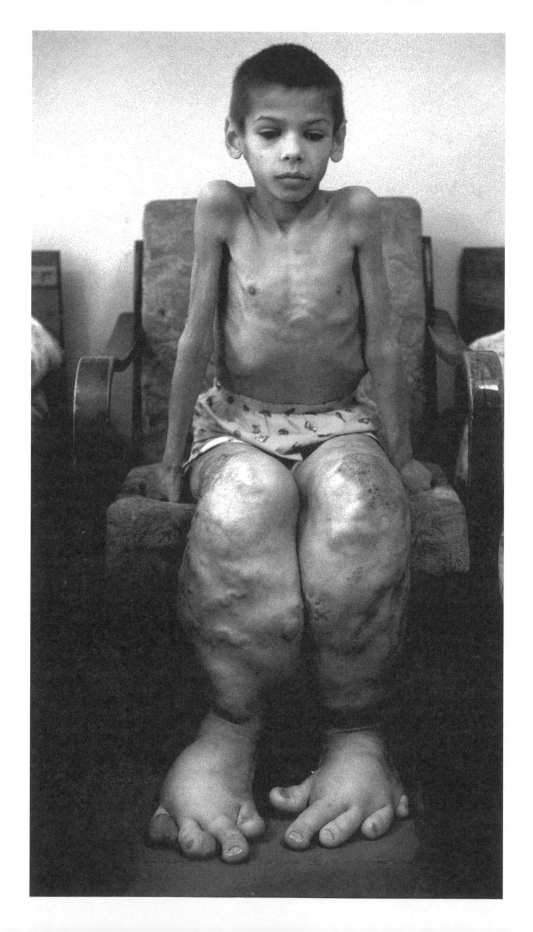

The greatest gift in life must surely be the gift of life itself. To be denied that must be seen as tragic.

We found that, privately, people would speak about thousands of women who wanted abortions because of foetal damage from radiation. According to the American Dr Robert Gale (who performed bone marrow transplants on some of the Chernobyl victims in 1986), about 100,000 abortions were carried out in the three affected regions in the first few years, abortions directly related to radiation. In 1993 it was stated that there were 200,000 abortions in Belarus and, when polled, the women gave Chernobyl as the main reason. There is no 'choice' involved. The power of the atom has taken away the power of women's right to make their own decisions. As a result, many of the women feel angry. Belarusian women are just like women the world over – they love their children and cherish the joy of giving birth. Yet for these women, the birth experience is all too frequently turned into sadness, and some women who find they have given birth to children who are ill or deformed leave their hospital beds and walk out, leaving their babies behind. Chernobyl is changing their joy into never-ending sadness.

In 1999, when we were working in Gomel, the main Chernobyl zone, the team and I came across two young doctors who were desperately trying to help families who were caring for terminally ill children at home without any medicine or hospice care. I asked to visit a family and found a nine-year-old boy, Slava, who had cancer of the liver and a tumour on his brain, lying prostrate on his parents' bed. He was incontinent and in and out of consciousness. His parents lived in a one-room flat. Slava lay on the only bed, while his parents and sisters shared the floor space. Slava had no painkillers, no incontinence pads or creams to ease his bedsores. His parents were doing the best they could, given their circumstances, but they had no money to buy what he needed. All they wanted for him was a death with dignity, a death free from the ravages of terrible pain. I was advised that this family was one of many. The doctors had no form of transport to travel between villages to visit the families. They begged us to give them two bicycles so that they could make their rounds. They also asked for some steroids, morphine and other drugs.

There and then, we decided to support these courageous doctors and families, and now there is a thriving hospice movement growing in Gomel. A couple of months after our visit, we donated not just the two bicycles, but a fully equipped ambulance, along with a six-month supply of everything they required to give other dying children some ease and comfort. Slava died just two weeks after our visit, but his legacy lives on, and in his memory this now-strengthened hospice movement will give to others what came too late for him.

The main long-term concern is the effect of continued, relatively low-dose exposure to radioactivity. Insufficient research on the effects of low-dose ionising radiation may well be the most important single reason for genetic damage, birth defects and cancer in the future. We are looking only at 20 years after the accident. Many scientists say it will not peak for another 50 years, so what does the future hold? One of the doctors I have spoken to told us that the Belarusian race is possibly threatened with extinction because their gene pool is now 'unclean'. The term 'unclean' is one I resent, and is a terrible indictment of today's world. The appearance of a whole range of cancers, neo-natal deaths, low-weight births and short pregnancies are all part of the legacy of Chernobyl. Because many of the rural hospitals have little or no access to incubators, the chances of the babies surviving are significantly reduced. One of the greatest joys I have experienced was during a convoy in October 1994 when we arrived at the Gomel maternity hospital to deliver an ambulance and a state-of-the-art incubator. I can still remember seeing a young mother clutching her newly born baby, waiting on the steps of the hospital as we struggled to get the incubator out of the ambulance. The doctors were rushing

us through the hospital corridors as we were hurriedly unwrapping the protective cellophane. Within minutes the incubator was plugged in and working. The mother

we had seen earlier came and placed her baby into the safety cocoon of the incubator. She started to cry, as did every

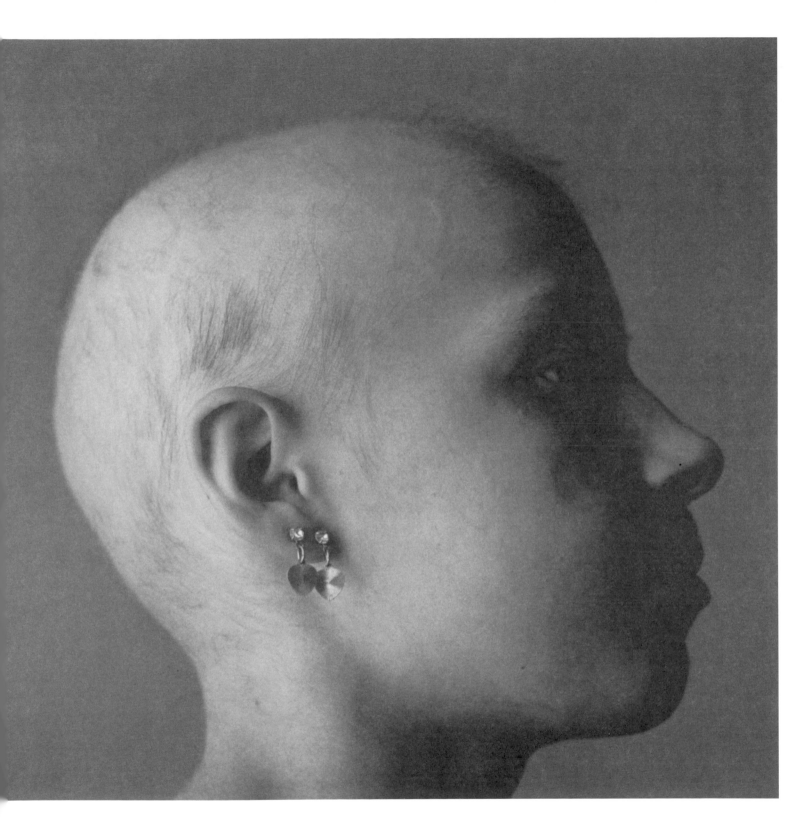

single one of us. If we had been a couple of hours later, her baby would have died. We saw a little miracle happen in front of our eyes!

In that same hospital the chief doctor told us he genuinely fears for the future of the Belarusian race. He looked at me and said, 'How would you feel if your race of people was threatened with extinction?' I couldn't answer. I began to understand what lay behind his question and was challenged by its implications. Emotionally and psychologically, is it possible to live with such a fear and threat? Radiation doesn't merely affect people in a physical way. It isn't a true picture just to consider the health effects without including the social, psychological and moral consequences of the impact of radiation – the destruction of a society's structure, people's fears for their children, the helplessness that emerges. The intangibility of the 'enemy' makes coping with it far less easy. Eventually a state of inertia sets in; people become immobilised and simply give up. Their powerless state is reflected in the number of suicides of people living in the zones.

The authorities try to dismiss people's fears and emotions by labelling it 'radiophobia'. Blame the victims. Nice and simple. The playing down of illness, both mental and emotional, is just one of the ways in which the victims are made victims all over again. No cognisance is taken of the interconnectedness of the

physical and psychological trauma caused by something like a nuclear disaster. They have packaged the 'effects' into nice, neat sums and equations on their computers. Nowhere in any of their calculations have they ever considered the genuine fears of the people. There is a widespread uneasiness among the inhabitants of the three

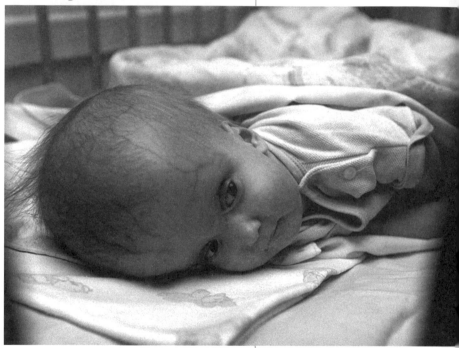

affected countries, a constant fear of being exposed to radiation and an awareness of the permanence of the threat to their health and lives. This has taken a severe toll and cannot be dismissed.

The potential genetic threat to the population over the coming decades may result in the emergence of a band of 'outcasts', particularly in the sphere of

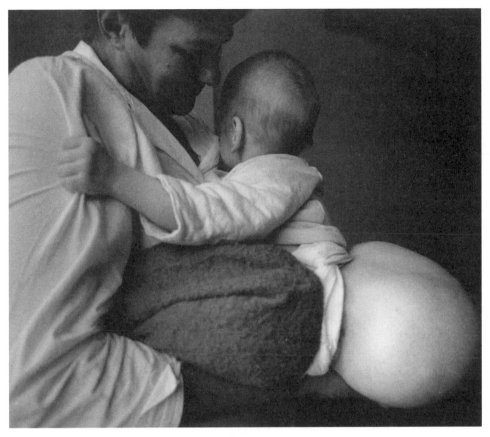

marriage and other human relationships. This happened after the bombing of Hiroshima and Nagasaki, when the survivors became known as the *Hibakusha* (outcasts). Many of them experienced discrimination, even to this day. Nobody wanted to marry them, employ them or be friends with them, for fear of contamination and because of ignorance. Some of them committed suicide, as they were unable to cope with the isolation and rejection – a parallel picture of the Chernobyl *Hibakusha*. Time and time again, people from all walks of life told us of their sense of being stigmatised and branded as 'Chernobyls'. This phenomenon has left people's confidence in tatters. The psyche of the nation, either consciously or unconsciously, is left confused and damaged.

As we walked through the corridors of the local Gomel oncology hospital, the now all-too-familiar bleak medical picture continued to be painted – the constant growth of blood diseases such as leukaemia, haemophilia and lymphoma.

The doctor showed us a graph of the cancer increase since 1986. The graph was shocking, but Dr Tatiana pointed out that even this wasn't a true reflection of the situation, because many of the children never reached the hospital for medical attention. I asked her what she thought about the future for the Belarusian people. She answered with a question: 'What happens when a sick child becomes a sick adult and has even sicker children?' She continued and gave the answer herself: 'The slow death of an entire race of people.' Dr Tatiana subsequently contracted breast cancer, which spread to her spine and other parts of her body. She died in 2004.

Dr Tatiana introduced us to a group of mothers and their children who were currently receiving treatment. These mothers were a courageous bunch of women who told us that sometimes the children died because they came from rural areas and had been diagnosed too late. The told us how they often had to carry their children great distances to get to the hospital because they had no car and public transport was limited and often unreliable. The mothers and doctors cried, but amidst their pain there was great strength. These women, encouraged by the doctors, had started to organise themselves and look for justice for their children. This was the first group we had come across in Belarus whom we felt had managed to cross from despair to empowerment, and had recognised their ability to seek their basic human right to protect the lives of their children.

Later, as we walked the wards, meeting and greeting the children, I thought about how we could possibly help to intervene in the lives of some of these brave children. I met 14-year-old Raisa. She was sitting alone, her beautiful long blonde hair loosely tied with a red bow. The doctor whispered that they had just diagnosed her illness as leukaemia. We were the first to know, even before her mother. As we moved from ward to ward, we met many more mothers and heard about how they had to share the tiny beds with their children. When a child was admitted to hospital, the mother would normally come to stay until the child recovered or died. This has put enormous pressure on families' already strained finances. Many of the mothers had to leave their jobs and had no guarantee that they would be re-employed later. Thus, the hidden effects of Chernobyl must also include the economic burden on the suffering families. For many of the families, the mother is the sole breadwinner, so when illness strikes a child, it affects the family both emotionally and economically.

The mothers had a tremendously strong faith in God. 'We are grateful for every day we have with our children,' one said. 'Each day our children survive is a precious gift from God.' The courage of these women and doctors left a marked impression. They said they sometimes saw aid convoys passing their hospital, but few stopped and entered. We resolved that one day soon a special medical aid truck, along with an ambulance, would arrive at their hospital.

As we left, one of the mothers gave us a special letter and asked that it be shown to everyone so that people might respond and help to save the lives of the dying children. Here is the text:

People of goodwill, help our children. With pain and hope we apply to you. We are mothers of sick children from the Gomel region. The common Chernobyl misfortune united all of us into one organisation, Children Haematology by name. There are children in our organisation who suffer from heavy blood diseases: leukaemia, lymphoma, haemophilia and others.

They need help.

We, the parents and doctors of the haematology department of the Gomel region hospital, fight for the lives of our children. Sometimes we have a success, sometimes not. It is a pity to say, but sometimes all our efforts end badly. The situation is extremely serious.

There isn't enough medicine to help all our sick children. And the children die…

Many of the parents have left work to give all their time and effort to these children. But we have no opportunities to take care of children-invalids who became sick after the Chernobyl catastrophe.

We need medicine, vitamins, ecologically pure product to feed our children. The children need to improve their health outside our sick country. Our organisation was founded to help our children. We have only few connections with foreign charity organisation. We need addresses of people who can help our children.

This is why we apply to all people of goodwill. Sick children wait for your help and in the hearts of mothers there can be a hope for the future. We shall be thankful for every support. We believe that kindness will save the world.

With this letter in my memory, 12-year-old Alesya Belyai from Klintsy, western Russia, came into my life in September 1995 when she arrived at Shannon Airport with a group of terminally ill children for rest and recuperation in Paul Newman's Hole in the Wall Gang Camp at Barretstown, County Kildare. (Hollywood actor Paul Newman has set up a number of adventure centres for terminally ill children.) I noticed how sad and shy she seemed. There was a quiet reticence in her being. I learned that she had leukaemia, but was in remission. Over the following

couple of weeks I watched her blossom and shine as she forgot her illness and painful road to recovery. She was initially in denial about what had happened and refused to talk about it, but little by little I got close to her and started to visit her and her family in western Russia whenever I was there with a convoy. The first time we met her parents, there were tears of gratitude and joy; they couldn't thank us enough for renewing their daughter's joy of life. I visited her with my friend Ali Hewson, and as we sat with the family in their tiny flat, their story unfolded of how they had to fight to receive any treatment for Alesya. They had been told to 'give up, accept her death, she cannot be treated'. Her parents would not accept this prognosis and sold their possessions to get her treatment in Belarus. The treatment cost them everything. Her grandmother sold her house and lived in a shed and her parents sold their wedding rings, all to try and save their daughter's life. So you can imagine their joy when she was in remission! Ireland was a 'God-sent blessing for Alesya, she became alive again!' said her mother. Over the following years we sent Alesya to the US for treatment and brought her to Ireland for recuperation. I watched her grow in confidence and she expressed her desire to learn English: 'I want to be a teacher for you, Adi. I want to speak English so I can help bring children like me to Ireland for their lives to be made better.'

Alesya reached her mid-teens, a pretty, attractive girl with an eye for style. But things changed. I started to receive frantic messages from her father that she wasn't well and that they had been told she had no hope. The cancer had come back, and with complications. I immediately organised for her to come back to Ireland under the great care of Mr Browne in St James's Hospital in Dublin. The medical team recognised immediately that she was seriously ill. Despite the odds, the team worked to save her. Alesya went through a lot of painful treatment and lost her lovely long hair again. Despite her ordeal, Alesya was conscious about her looks and asked if we could get her a wig. Not just any old wig, she told Ali, 'I would like a Jennifer Aniston wig please!' And yes, Ali bought her the Jennifer wig. Alesya was in and out of hospital over a period of months, and when she was in her well periods she had a wonderful time. We brought her to restaurants, shops (she loved clothes shopping!), she got her tummy pierced, an ankle chain and all the latest gear. None of us ever thought she wouldn't get better, but one day in June 2000 the consultants told me that the fight was over; they couldn't save her life, that we should send her home to die. As we reeled from this news, we had to prepare for the worst. Her mother, who had arrived in Ireland to be with her, refused to accept it and wouldn't discuss what we should do. I brought Alesya and her mum to the airport at Shannon, where

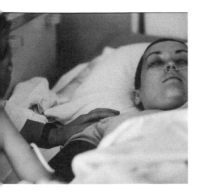

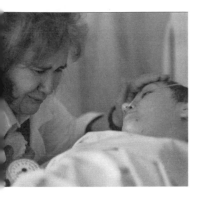

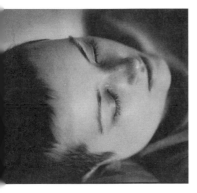

Alesya's last moments

we said our goodbyes. I promised I would visit in the autumn. I asked Alesya to 'hang on' until I got there. She promised she would. On her return she became very ill, but nothing could be done at the local hospital in western Russia because they had no medicine. We organised for her to be moved to Belarus and to receive the best of medical care in the cancer hospital in Gomel, under the expert care of Dr Tamara. This was highly irregular, but was allowed as a favour to our organisation for all we had done. Alesya deteriorated quickly, slipping in and out of a coma. All this time her mother never accepted that she wouldn't pull through. After all, Alesya had been stricken with leukaemia when she was 11 years old and for six years had battled with her illness. She finally died just after her seventeenth birthday. She died in the arms of her mother, who whispered, 'Daughter, daughter…my beautiful flower. Stay with me…stay with me…my little star, don't leave me…don't leave me…my love, my life.'

When Alesya was three she had been bathed by the black rains of Chernobyl.

But sometimes there can be a happy ending. In February 2005 I was invited to a number of high-level political meetings by the Belarusian government. We saw this as a significant request and put a lot of effort into preparing for it. Project business consultant Sile Byrne and I had a hectic three days of meetings not only

with top government officials, but also with the press and non-governmental organisations. It was the first time in all my years of visiting Belarus that I didn't have any involvement with the physical reality of Chernobyl as we moved from one comfortable office to the next. Indeed, I felt remote from it. But all that changed just hours before we left while holding a meeting in the Project's tiny Minsk office. Something happened that was to change the meaning and purpose of the trip.

While talking with a local Chernobyl organisation, I noticed someone quietly standing at the door holding the hands of two children. When I looked again, I could see immediately that the two children were sick. They were both sickly pale, thin and with solemn faces. The three of them framed the door. I invited them to come in and sit down and the mother of the children immediately started to speak with speed and urgency. I soon learned why. Angelica, the mother, had seen us on television and had spent three days going around Minsk from one government department to another trying to find us. Her first sentence to me was, 'My baby, Lisa, is dying. Please save her. You are our only hope.' She went on to tell us how Lisa, six years of age, couldn't be helped in Belarus and that she had been told to 'accept your daughter's death; take her home to die'. A raging torrent of words poured out, telling us of her life and

67

that of her children. She was very poor and couldn't afford to take her children abroad for treatment. She begged us to help. Angelica said that the most common words used in her children's lives were 'death' and 'dying'. I became conscious that both children could hear everything she said, so to distract Lisa I asked her to draw something with a pen and paper. She immediately responded and was busy for some minutes until I asked her to tell me what the lines she drew meant. Lisa replied, 'I write to God every day and this is my letter today.' I asked her, 'And what do you say to God?' She replied, 'When will it be over? When will I die?'

Silence followed, interrupted by Angelica's quiet sobbing. While I wouldn't make any false promises to help, I knew that I couldn't turn my back on this little girl's plight. I immediately phoned a Belarusian photographer friend, Anatoly Kleschuk, and asked him to come and take some pictures so that I could release the story and make an appeal on my return home. I realised just at that moment that

the deeper purpose of our being in Minsk was to try and save this little girl's life.

We arrived back to Ireland and within two days had an appeal in one of our national papers, followed by a television appeal on TV3. Between both appeals we raised enough money for Lisa to be sent to Moscow for life-saving treatment. As I write, preparations are being put in place for the journey to Moscow for her life-saving surgery.

Chernobyl is a continuing human drama, one that requires a complex array of responses, providing the gamut of services from diagnosis to treatment to psycho-social support. Left alone, the three affected countries have to deal with unstable economies and weak health services. Their future is grim. The world needs to know the truth about the consequences because there could be other potential Chernobyl disasters in the future. We owe this to humanity in order to prevent others from sharing the same fate as the innocent men, women and children of Chernobyl.

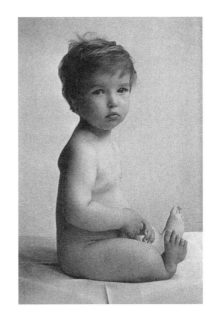

Anna Gabriel at the home for abandoned babies in 1993

Anna Gabriel at 13 with Adi and Ali, her godmother,
proudly showing off her new prostethic legs

Chernobyl Heart is the name given to a previously unheard-of condition. I first heard of it in the Children's Regional Hospital in Gomel in 1994. I was in the corridor when I heard the patter of many small feet running after us. A voice called out something to me in Russian. Stopping, I turned around to be faced with about a dozen children from the last ward we had visited. I sought the face belonging to the voice. It was one of the children whom the doctor had pointed out as being extremely ill. It took the translator some moments before he could compose himself to translate. The little boy, Vitaly Gutsez, nine years of age, had said, 'Please take me to your country, I will die if you leave me here.' My head wanted to disbelieve the stark words of the child. My heart told me it was the truth, a truth confirmed by Dr Russokov when he told me that Vitaly had what they term 'Chernobyl Heart', where caesium-137 causes heart dysfunction, causing children to have severe heart damage, even heart attacks and strokes. He continued, 'We have no medicine, no equipment, Vitaly is doomed to die, we don't know when but we know he will die for sure.'

Vitaly was already in a deteriorated state. His skin had a blue/purple tinge to it; his lips, tongue, eyelids, palms and feet were black; and the tips of his fingers and toes were seriously clubbed. I felt powerless confronted with the reality of this shocking news. I recalled other children I had known who had had little time to live. The children died, in pain, forgotten. This time it would have to be different. I couldn't let Vitaly slip away. The face of this young boy would haunt me forever if I did nothing. As we walked down the corridor I began to hatch a plan.

In the same hospital a seven-year-old girl named Evgeniya was in a room alone with her mother, Luda. The doctor told me that they couldn't even diagnose her problem. Looking at Evgeniya was hard, as all her veins were enlarged and distorted. Her body was swollen and she looked more like an old woman than a child. Her immunity was low and she ran constant high temperatures. She had a forlorn, resigned look. Luda stood at the bottom of the bed looking anxious and pale. She had kept a vigil at her daughter's bed on and off for six and a half years. She was tired and worn out and had no idea what the future held for her daughter. As long as there was no diagnosis, there could be no treatment.

Like every hospital we visited in Belarus, the doctors and staff were highly qualified and dedicated to their patients. The grim reality of little medicine, poor diagnostic equipment and a continual increase in the number of patients made their job almost unbearable. I have to admit that on leaving that hospital I felt relieved, but I couldn't stop thinking of the words of the dying boy, Vitaly. We agreed to make some attempt to do something. I wrote a

70

Above: Vitaly and Sasha before their treatment
Below: The boys after their treatment

fax message to our office in Cork advising them to contact our local hospital for help. The response was swift and definite. The Mercy Hospital in Cork had responded, thanks to the plea from Sister Fidelma on our behalf. They would accept Vitaly for treatment. When I phoned Dr Rusakov he was a happy man, but then he said, 'Adi, this is good news, but what about the others?' I responded, 'What others?' Dr Rusakov continued, 'You remember the girl, Evgeniya, and the little boys Sasha and Artem? You took their photographs! Well, they need treatment too; otherwise I am sure they will not survive.' What could I say to that? Therefore, instead of organising to bring one child, we now had the task of bringing another three, which meant a lot more persuading and more paperwork, but there was no option.

In the following few weeks there were many anxious moments when it seemed that it wouldn't be possible. We ran into practical problems given that it was Christmas week and most of the government offices we needed in both Ireland and Belarus were shutting down for a number of days. We had to work fast. However, the main problem was a logistical one, related to the condition of all four children. The doctors in Gomel feared they wouldn't survive the flight, as

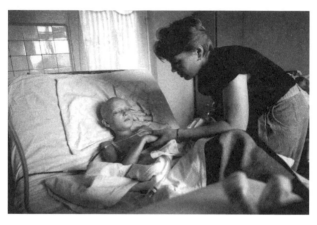

the high altitude would affect their weakened hearts. Then there was the question of the long eight-hour journey in freezing cold conditions to the airport, a two-hour wait for a four-hour flight followed by a further three-hour journey by ambulance to Cork. The parents had a tough decision to make. Would they take the risk of sending their children on such an arduous journey to a foreign country they had never heard of, with people who were total strangers, and run the risk that they might possibly die along the way? Alternatively, should they keep them at home and accept and prepare for their deaths? I understood their dilemma, as well as the responsibility it would place on our shoulders if they agreed to let us take them.

No airline would take them once we explained their condition, so we had to find another way. After a few dozen phone calls on Christmas Eve 1994, I eventually got an Icelandic businessman to agree not only to lend us his plane to airlift the children but to also pilot the aircraft himself. It was only thanks to the dedication of Dr Rusakov and his staff, along with our own embassy in Moscow and our Departments of Justice and Foreign Affairs, that we finally got the paperwork finalised. The rules were made flexible to make it possible. The Irish embassy even stayed open late on the day they were due to close for Christmas in order to facilitate Vitaly's mother, who had to make the frantic journey of 14 hours by train to complete his paperwork for a visa. They did it because their hearts were touched by Vitaly's plea to us in the hospital corridor. On New Year's Eve, I remember going to bed early as I was leaving for the airport at four a.m. the following morning to join the second pilot, John Backhouse, in Liverpool. John was flying us in his own small aircraft to Iceland to pick up Thor Magnusson, the Icelandic businessman, and head off to southern Belarus.

Arriving at Gomel Airport in what seemed like the middle of the night was a disorientating experience. It was freezing cold – minus 10 degrees – as we stepped out of the snug aircraft. After a restless night we finally arrived at the hospital to pick up the four children. They shyly watched me approach. They were unsure of how to behave until Vitaly 'The Brave' (his new name!) threw his good arm around me and landed a big kiss on my cheek. Chatting away to the children,

holding as many hands as I could, I eased my way into what would become a long-term relationship. These children were being entrusted into my care. It was an overwhelming responsibility to think about, but at the same time I felt it was a tremendous recognition of the trust and faith of the Belarusian people in the Irish.

One of the mothers stepped forward and said, 'We understand that the next time we see our children may be in a coffin.' With these words she had released me from my fears that if we couldn't save their lives that they would understand.

The flight was memorable! None of the children had ever been in an aircraft and were fascinated by everything, and despite their upset at leaving their parents they laughed and clapped and sang as we travelled through the clouds. The next five and a half hours had difficulties, particularly for Evgeniya, who was in constant pain. We had to fly at a fairly low altitude because of the possibility of heart failure. I spent my time distracting the children with tales of the adventures that lay ahead in Ireland. They were fascinated that we would be on an island and by the sea. None of the children had ever seen the sea.

Both Vitaly and Sasha had to spend most of the journey with oxygen masks on and there were many tense moments when we thought that either of them would go into heart failure. Dr Irina expressed her concern to me that we might have to make an emergency landing. Our pilot kept radioing various airports en route to alert them to the possibility of an emergency. The journey seemed to take forever, but finally we could see the lights of Cork Airport ahead. As we taxied down the runway I could see a huge crowd of dancing figures ahead of us. Mary Aherne, Seán, Sister Fidelma, the Knights of Malta, volunteers, airport staff, and of course the media, were all cheering and waving. As the children were gingerly carried off the aircraft, there was a special moment of hush amongst the crowd. The beauty, the vulnerability, the specialness of the children moved the crowd. The children won the hearts of the nation from the second they were carried off the aircraft. As the doors of the ambulance closed I felt a whoosh of relief that the miracle had started.

The following day dawned to a blaze of wonderful photographs and television coverage. The children were now famous! Irish children started to send the Chernobyl children get well and good luck cards. There was also a constant stream of adults with their families coming to bring gifts to our children. Almost immediately their room became like a toy shop. The children were given a special room to themselves and quickly became the centre of attention with all the other young patients. Until this point we hadn't dared to think beyond getting the children to Ireland. With that part of the mission completed, we were faced with the question of what we should do now. The children would need specialised treatment which wasn't available in Cork, so we had to set about the delicate process of getting the children accepted into a hospital in Dublin. Yet again, everything worked in our favour, and with the help of Dr Seamus O'Donoghue from Mercy Hospital, we got permission to send the most urgent two, Sasha and Vitaly, to Our Lady's Hospital for Sick Children in Crumlin. The day we left was the Russian Orthodox Christmas Day, and prior to our departure we managed to get a patient from the men's ward to dress up as Santa and to give the children their surprise presents. The children were genuinely not expecting anything, and the look of surprise on their faces was something else!

In Dublin we were met by my sister, Len, and my friend Eoin Dinan, who were dressed up in Christmas suits. Despite the serious purpose of our journey to Dublin, we kept the spirit of Christmas going.

Settling the children into yet another hospital was no problem, as their lives had already consisted of many hospitals. But Sister Fidelma and I found it hard to leave the children. We had even declared them 'honorary Corkonians'. Parting from them was harder for us than for them. The next time we would see them they would both have undergone major surgery, and they wouldn't have the smiling faces that we now saw. But leaving them in the special

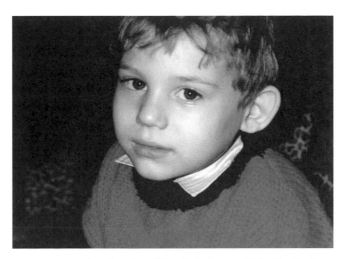

Artem

Evgeniya

hands of Dr Oslizlok and Mr Duff, along with their medical team, was reassuring. If anyone could provide us with a medical miracle, I was sure that these doctors would be the ones.

Moving Vitaly and Sasha took its toll on Artem and Evgeniya. The little family that had grown together was split up, and the two left behind missed the others dreadfully. We had our first bout of loneliness when five-year-old Artem took out a little notebook with his mother's phone number in it and brought it to me with tears rolling down his pale cheeks, saying, 'Mama, Mama.' He told Tanya that his heart would break if he didn't speak to his mother. We promised that we would try and phone in the corridor. Armed with massive amounts of 50p and 20p coins, we eventually managed to connect mother and son. He cried and cried and told her to come '*zaftra, zaftra*' ('tomorrow, tomorrow'). Artem took out his notebook

every day and shed more and more tears, but thanks to the love of the hospital staff and the other children and their parents, he survived his stay in Cork away from his parents with relative ease and comfort.

Eventually Artem was taken by Sister Fidelma and Tanya to Crumlin Hospital, where he underwent extensive testing. Finally he was returned to his 'home' to Gomel. The doctors have given him the 'all clear' and are quite sure that he will live a long and healthy life.

It's harder to write about Evgeniya, the quiet, shy seven-year-old who had to lie on her bed, unable to move because of her severe pain. Day after day I called to see her, and the more I saw her, the more I admired her. Despite all that suffering, she always managed to smile and to blow me a kiss when I would get up to leave and say '*spekone noiche*' ('good night'). The doctors started by reducing the steroids which had bloated her body so grotesquely and by detoxifying her of the cocktail of drugs that had been pumped into her over the short years of her life. The withdrawal wasn't easy. It increased her pain and made her even less mobile. Every so often Evgeniya would ask questions like, 'Will I be getting better soon?', 'When I grow up will I be able to have babies?', 'Will the babies be like me?' Her constant probing questions had to be answered without giving her the complete truth. We couldn't fathom what could be done for her, but we had to believe that everything possible had been done and that it would at least improve her condition and at best cure her completely.

In Crumlin things moved with great swiftness and expertise. The team of medics decided to operate on Vitaly first and later on Sasha. I visited the boys only once before the operations, and it was a day I shall not forget too easily. A couple of days before the operation, the hospital administrator, Paul Kavanagh, phoned me to say that he and the medical team felt it would be of primary importance that the two mothers of the boys came in advance of the operations. The logic was that with their mothers present, the children would relax much more easily beforehand and that in the crucial days in intensive care afterwards the mothers would be a key factor in the boys' recovery. I knew he was right and that, of course, the boys needed their mothers most of all, but I cringed at the thought of the mountains that would have to be moved to make it happen. As there were only two flights per week, it left us with only three days to organise the passports, visas and tickets. Not an easy task, when you consider that it normally takes months to get a passport in Belarus and even longer to get the paperwork for a visa sorted out.

But Dr Edourd and our translator, Luda, helped to make a special plea to the local authorities, and within 24 hours the two women had their passports. That only left the Irish visas. True to form, our embassy was willing to waive the visas and to give a letter instead. But the mothers were so terrified that they wouldn't be allowed out of the country without the actual stamp in their passports that they went off to Moscow on a 14-hour train journey to our embassy. They were greeted immediately by Julian Clare of the Irish embassy, who gave them their visas within minutes. So the two women set off down to Belarus to prepare for the journey to Minsk Airport.

From there these two women, Tamara and Nazedha, left their country for the first time in their lives and set off to Ireland. They were greeted at Shannon Airport by our volunteers, who drove them to Dublin. In the meantime I managed to get the train to Dublin, so I was there to greet the women, who arrived, appropriately enough, by ambulance. The excitement was marvellous as the women arrived at the hospital, to be greeted by Paul and Eoin with flowers, smiles and kisses.

Immediately Tamara and Nazedha asked about the boys and when they would be able to see them. As we waited at the lift I could feel both the mothers' emotion and anticipation. Eventually the doors of the lift opened, and there were Sasha on a bed and Vitaly in a wheelchair. They were wheeled out to greet their mothers. The boys cried, the mothers cried – in fact, it appeared that everyone shed a tear or two! It was hard not to get engaged in such a moment. The mothers were overwhelmed, but nothing, not even their extreme tiredness, could spoil their reunion with their boys. That day is a day that I shall always treasure.

As the mothers settled into hospital life in Ireland, the medical team prepared them for what lay ahead for their sons. Both Tamara and Nazedha were pale and frightened but were constantly reassured by Dr Oslizlok and Mr Duff that everything would be all right.

On 18 January at 8.15 a.m., Vitaly's operation started. He was on the operating table for many hours under the experienced hand of Ireland's best-known heart surgeon, Mr Maurice Nelligan. Vitaly subsequently spent time on the heart-bypass machine. Tamara held an eight-hour vigil outside the operating theatre and practically collapsed when the operation was declared complete and successful. Her only words were, 'My son has come back from the dead. Thank you! Thank you!'

Vitaly was put in intensive care. It seemed like a million tubes were coming in and out of his body, but he had survived. Every second, every minute, every hour, every day that he now survived would be a strengthening indication that he would pull through. No one dared ask about his condition, but we all knew that his miracle had happened, and that there was every possibility that he would make it. Over the next week, little by little Vitaly

started to come round. The blueness of his skin and lips started to fade and he became more and more like any other little 10-year-old boy. After eight days he left intensive care in his wheelchair and smiled and waved at everyone. Sasha looked relieved, as he could see that Vitaly had not only survived but looked wonderful! Seeing Vitaly so well and healthy was a great encouragement for Sasha, who would have his operation on 25 January. Since Sasha's operation had been postponed due to a slight infection, he had been fretting in anticipation of what lay ahead, and now he felt reassured and ready to face his life-saving operation.

In the middle of all this I received a phone call from the Foreign Minister of Iceland. He had seen the Icelandic news items about what had now become the famous 'mercy rescue mission' and wanted to know what Iceland could do. I agreed to fly out to talk with their Foreign Ministry and Department of Health. The Icelandic Minister for Health announced on television that he would invite other seriously ill children from Chernobyl to go to Iceland for treatment.

Two weeks after Vitaly's operation I collected him and brought him back 'home' to the Mercy Hospital in Cork, where he was put in the same room as Evgeniya. As he was recuperating we asked our rota of Chernobyl supporters to take him on short trips around Cork. He was taken to see the sea, which he had never seen before. His face lit up with delight and his joy was passed on to all of us.

Early on the morning of 13 February I was awakened by a frantic buzzing of our doorbell. Half asleep, my husband and I looked out of the window and saw two strange men in our front garden. Bewildered, I crouched at the top of the stairs while Seán answered the door. One of the men identified himself as the night porter from the hospital. He said, 'Something is wrong with one of the children and you must come immediately.' Still not quite awake, we pulled on some clothes and drove to the hospital. I felt as if it was part of a dream. We were met by the matron, who told us that Evgeniya had just died. In a daze of disbelief we went to the ward. A nurse motioned us behind the curtain that surrounded Evgeniya's bed. She lay there, in what looked like peaceful sleep. She had died just minutes before our arrival. The nurses on duty were distraught as they told us of her last moments in life. It hadn't been an easy death. Finally she had given up the struggle to live, and so her last 10 minutes had been peaceful.

Seán and I were joined by a tearful Tanya, who had been Evgeniya's lifeline and solid friend. The three of us held each other for support, each of us locked into our separate but shared grief. We helped to take Evgeniya to the mortuary, where we sat and cried and intermittently touched her body. She looked so beautiful.

We were in complete shock and disbelief. No one had expected her to die. Tanya and I had only just helped to wash her hair on Sunday, in preparation for her mother's visit four days later. We had laughed and chatted about what her mama would do when she saw her looking so pretty in the bed.

When Evgeniya's mother, Ludmila, finally arrived in Ireland the following night, she was so overwhelmed by her sadness and exhausted by a new and unfamiliar world that she was incapable of any communication. We brought her straight to the quiet room where Evgeniya was laid out. She moved to the coffin and paced around it. She touched every inch of her child's body, all the while uttering soothing, gentle words of reassurance.

Later that night we talked and cried together and shared photographs of Evgeniya. I told her all about Evgeniya's time in Ireland and how everyone loved her. Ludmila told us about a dream she had had in the early hours of Monday morning: 'I dreamed I was in the hospital in Ireland and saw Evgeniya running around the corridors. I called out to her and said, "Evgeniya, what about your pain?" She turned to me, and smiling, said, "Mama, there is no more pain."' Ludmila believed that this dream was her daughter telling her that she needed to be free from suffering and was allowing herself to go.

In the days that followed we slowly pulled ourselves together. Two weeks after

Evgeniya's death, Vitaly and I set off for Iceland. As we flew over the ocean I thought about how much I needed to recapture the joy of the other three children, whose lives had been saved, and Iceland was to give us that opportunity. We spent a week as special guests of the Icelandic people, who welcomed us with open arms. Vitaly was a celebrity there, as everyone had been following his progress since the day we had airlifted the children to Ireland. During our week we met the president, Vigdis Fingvogadottir, a woman well known for her love of children, and she gave Vitaly her complete attention.

On our return to Ireland, Vitaly was finally discharged from hospital into our care and eventually returned at the end of March 1995 with the happy prognosis of at least another 30 years of life ahead of him. Sasha had a further two operations and eventually returned home in April 1995.

In late April 1995 I made a quick visit to see the three boys and their families in Gomel. Within minutes of our arrival I was surrounded by everyone and we greeted each other as old friends. There was much speech-making, with strong words of gratitude to the Irish people. The most special words for me came from Vitaly's father when he stood up and said, 'I can die now because my son lives, thanks to the people of Ireland.'

Vitaly and Sasha are now two strapping teenagers with their whole lives ahead of them. They still come to Ireland for annual check-ups with Dr Oslizlok and stay with their original families.

Those first four children taught me a lot. Mainly they helped me to go beyond sending humanitarian aid and to get to know the real suffering by touching the death and the pain, to take the risk and to go deeper. Because they showed me how to be brave, I've taken many similar risks and in the name of Vitaly, Sasha, Artem and Evgeniya, our organisation developed our Life Saving Operations Programme, under which we have intervened and saved the lives of hundreds children.

The story of these four children introduced us to the little-known condition of Chernobyl Heart as far back as 1995. Since that time, unfortunately, we were to find many more children with the same condition.

In 2000 I walked the same corridor in Gomel where I had found Vitaly Gutsev. With these memories on my mind, I felt anxious as I entered the room where Vitaly had lain. In his bed I saw another little boy lying listless, looking extremely ill. At the end of his bed was his mother, Tatiana. When our friend, Dr Irina, introduced me, I asked about the little boy, learned that his name was Alexei Cherpernoi, and that he suffered from the same Chernobyl Heart condition as Vitaly. Tatiana started to cry bitter tears; Alexei just lay there, motionless, following us with his beautiful eyes. All I could say

Alexei Cherpernoi (1994–2000)

to Tatiana is that we would try to help, and that these weren't empty words. I went on to say that if she would allow us, we would try to intervene.

On leaving the hospital I decided to make a call to the CCP office in Cork and dictated an emergency fax letter to a medical friend, Mr Michael Earley in Temple Street Hospital. I begged him to help. I told him how shocked I was at Alexei's condition. Michael Earley said the magic word. The frantic move to get Alexei and one of his parents to Ireland began. The next time I met Alexei and his father, Nicolai, was at the airport. We were believing in 'where there is life there is hope'.

When the medics at Temple Street examined Alexei that night, they knew things were extremely serious. Alexei was swiftly moved to Crumlin Hospital under the care of the same team, led by doctors Duff and Oslizlok, who had saved Vitaly. He was in exactly the same ward as Vitaly, St Brigid's ward, and I felt this was another good omen. Over the following weeks Alexei wooed the hearts of everyone. On days where he was feeling stronger, his personality shone through – what a sense of humour and sense of play! Alexei brought a special light into the lives of those who met him.

Around his hospital bed I learned a lot about Alexei from Tatiana and how he loved animals and picked her flowers. She said he was a 'thoughtful little boy'. When

she had arrived in Ireland to be with him, despite being bedridden and hooked up with all sorts of tubes and equipment, his first words to her were, 'Are you tired, Mom, after that flight? How is my baby sister? How are the animals?' He often told his parents when to go and have a rest because 'you must be tired from looking after me'. He was acutely tuned in to his parents' emotions, and often when Tatiana was sitting by his bed, deep in thought, he would say, 'You miss baby Kristina, don't you? I wonder how she is without us, she is getting bigger every day, I suppose.' He missed his little sister a lot and constantly asked questions about her, especially on the days when they phoned home. He was looking forward to Christmas and the New Year. One day he got a small Christmas tree and he spent a lot of time decorating it. A few days later he wrote in the letter he gave me to give to Santa Claus that he would like a set of three tiny racing cars and a 'Tom and Jerry' video. But he dreamed of going home for New Year. He told us where they used to have their New Year tree and of the toys he was going to use to decorate that tree. His only regret was that Kristina was maybe still too small to enjoy the tree. He asked us to buy a doll for her, but he didn't want his mom to know about his request as she would be annoyed with him for asking!

We quickly learned that Alexei was in such a deteriorated condition that he was almost gone beyond saving. Our medical team battled nonetheless to try everything medically possible. Unfortunately, the odds were against them, as Alexei had been left for so long without the appropriate intervention.

Alexei deserved to live life to the fullest, but that wasn't meant to be. He died shortly before Christmas in the loving arms of his mother and father at the tender age of six. The courage that Vitaly had shown me helped me to deal with the loss of Alexei and learning to understand that in reality, we cannot save all the children. Being with Alexei was a privilege, a gift. No matter how short his life was, he left a considerable trace in my life and in the lives of all of us who were lucky enough to know him.

Today in Belarus alone over 7,000 children await treatment for Chernobyl Heart, in addition to an estimated 800 to 1,000 newborn babies being added to this list every year. Without intervention, the children on the waiting list have a short life span of between three to five years. However, in the summer of 2003 our American sister organisation, Chernobyl Children's Project International, got in touch with top US cardiologist Dr William Novik, who travels the world operating on children with cardiac problems. Between ourselves and our sister organisation we agreed to fund a cardiac team to fly to Belarus and start tackling the long, daunting list of Chernobyl Heart victims. The first mission we sponsored was in October 2003, and since then we have sponsored many more trips that have saved the lives of over 200 children. While we are still only making a dint in the overall figure of children suffering, we feel we have to concentrate on saving as many lives as we can. Thanks to our work with Dr Novik and his team, we have gone from being able to treat only one Chernobyl Heart child a year to almost 100. Our Chernobyl Heart Programme, as well as the impact of saving so many precious lives, has the additional task of providing 'side by side' training of local specialists. Eventually the objective is that all children will be operated on in plenty of time by local surgeons.

Imagine – the life of a child can be saved by our intervention. It's humbling to see that for as little as $1,000 we can make a miracle happen. The children put their hearts in our hands, and one by one we try and make a miracle happen.

Since the accident, Central and Eastern Europe have undergone momentous political changes. The USSR no longer exists. Chernobyl is now the responsibility of the respective governments of each of the affected countries, but the fallout from Chernobyl continues to kill and mar the lives of millions. Despite all the words that have been written about the accident, little has changed for the better. In fact, in many ways the situation is getting worse.

For the first four years after the accident, we went from no information to the general public, to limited and selective information coming from narrow scientific circles. Only now is the chilling reality starting to unfold.

We aren't finished with Chernobyl. The scientists admit that the cement 'sarcophagus' which encases the damaged nuclear reactor is now cracking open and leaking out lethal doses of radiation. In 1988 Soviet scientists announced that the sarcophagus was only designed for a lifetime of 20 to 30 years. Holes and fissures in the structure now cover 100 square metres, some of which are large enough to drive a car through. These cracks and holes are further exacerbated by the intense heat inside the reactor, which is still over 200 degrees Celsius. The sarcophagus's hastily and poorly built concrete walls, which are steadily sinking, act as a lid on the grave of the shattered reactor.

The scientists now agree that this sarcophagus will eventually collapse, and when it does there will be an even greater release of radioactivity than in the initial accident. Only 3 per cent of the reactor's lethal material was expelled in the initial accident, leaving 97 per cent within the unstable sarcophagus. Twenty thousand tons of concrete floor is about to collapse into what can only be described as a mix of radioactive lava and dust, which resulted from the dropping of tons of sand in the early attempts to put out the fire, formed by the fusion of molten fuel, concrete and dust. This tomb, which was meant to last forever, began to deteriorate in the first five years. The pillars and beams supporting the building which contains the damaged reactor are in serious danger of bursting. If this is allowed to happen, the consequences could include the crashing of debris right through the concrete sarcophagus, or rubble could lunge into Reactor 3, which is right next door. This could trigger a core meltdown, which would in turn send another radioactive plume into the atmosphere that would blow over all of Europe and beyond. There are 740,000 cubic metres of lethally contaminated debris inside the sarcophagus, which is 10 times more than was previously thought. Inside the tomb there is an uneasy sinister scenario in which no one can predict the outcome if left unchecked. Locked inside lies 30 tons of highly contaminated dust, 16 tons of uranium and plutonium and 200 tons of radioactive lava. Because it isn't sealed, the

THE SARCOPHAGUS

'The next Chernobyl will be Chernobyl itself.'
Professor Chernosenko, Russian scientist

84

rain and melting snow pours through, causing corrosion. The weight of 3,000 cubic metres of water lodging each year further adds to the possibility of the roof caving in. The result of the water and dust mixing is a dangerous radioactive 'soup'. When the building became highly radioactive, the engineers were unable to physically screw down the nuts and bolts or apply any direct welding of the sarcophagus. This work was done by robotics. Unfortunately, the result is that the seams of the building aren't sealed, thus allowing water to enter and radiation to escape on a daily basis. The problem of controlling the water and dust inside has never been resolved. The nuclear fuel has to be extracted, controlled and buried.

'The sarcophagus is so porous that radioactivity escapes each day,' says the former director of security services from Ukraine. He continues, 'We don't even have the ability to measure the amount. If we could see the radioactivity there would be a cloud of smoke over the sarcophagus. What would happen if the cover collapses depends on the wind. In September 1996 we recorded the last atomic chain reaction, but it's very possible that something is happening now. We just don't know.'

His bleak statements are corroborated and further expanded on by leading Russian scientist Professor Alexei Yablokov, a member of the Russian Academy of Sciences and former adviser to ex-president Boris Yeltsin. He states that on several occasions experts have seen a luminescence characteristic of chain reactions inside the sarcophagus. He says, 'If it collapses, there will be no explosion, as this isn't a bomb, but a pillar of dust containing irradiated particles which will shoot 1.5 km into the air and will be spread by the wind. This could be comparable in its scale with the disaster itself.' Professor Yablokov, in his capacity as president of the Centre for Russian Environmental Policy, states that nuclear reactions are taking place spontaneously inside the sarcophagus as rain and snow fall onto the unspent fuel. Further verification of Professor Yablokov's statements comes from the Head of Information at Chernobyl, Julia Marusych: 'This type of project has never been undertaken before and no one knows for sure if it will be effective enough to contain the radioactivity or what will happen in a hundred years' time. It has taken 17 years to get an agreement to rebuild the crumbling sarcophagus at the cost of over $1 billion. I just hope it doesn't come too late. I am reminded of the chilling words of the Russian scientist, Professor Chernosenko, when he said, "The next Chernobyl will be Chernobyl itself." I pray to God the new sarcophagus will be built in time to avoid this nightmare prophesy.'

Other experts also agree that it will take at least 100 years to complete the Chernobyl clean-up. According to Dr John Large, a leading British independent

nuclear safety expert, Chernobyl's debris will be radioactive for hundreds of thousands of years and must be treated and buried in shallow graves as an urgent priority. Meanwhile, the Chernobyl time bomb is ticking away as politicians argue about who will pay. But what price will we pay if nothing is done? There is little doubt that the sarcophagus is sinking into the ground. 'There are simply no proper foundations,' says German radiobiologist Professor Edmund Lengefelder. This giant concrete shell is falling apart and casts a dark shadow over the already gloomy picture. It's a graphic symbol of the nuclear monster that isn't willing to lie down and die.

The Chernobyl disaster represents the first large-scale 'experiment' in the management of a nuclear crisis, and it has failed miserably. It has had a social impact unparalleled in human history. People's view of nuclear technology has been changed forever. The accident that scientists were so convinced 'could never happen' has marred the lives of millions and caused the devastation of what was formerly known as 'the breadbasket of the Soviet Union'. The scientists were paralysed into inactivity by their own deception. It was as if to admit to the accident was an admission of defeat. The scientific community must now recognise that the human body's response to radionuclide pollution is far more serious than ever imagined by the nuclear industry. We haven't developed resistance to natural or man-made radioactivity. Nuclear science has even created new radioactive elements such as plutonium and caesium which had never before existed on the living planet.

But there is hope, finally. In 1998, with the help of the European Bank for Reconstruction and Development (EBRD), a stabilisation programme was completed which included securing the roof beams from collapsing. In 2004 all preparatory work for a new sarcophagus was well advanced and in 2005 engineers moved towards completing plans for what may be the largest moveable structure ever built, a 20,000-ton steel shell to replace the failed sarcophagus. If these plans are successful, the construction will be the height of a 35-storey building. Inside,

robotic cranes and, where possible, live workers will then begin the delicate job of prying apart the wreckage and removing the radioactive materials. This construction will not be completed until 2011 and at a cost of over $1 billion. And then the world will wait. The shelter is designed to keep water out and dust in for approximately 100 years, or for as long as it takes the Ukrainian government to designate a permanent storage facility and dispose of all the radioactive material.

It's a dangerous task, and with no previous experience of such a project to draw on, this work will have to be carried out in the most toxic environment in the world. Excavation is complicated and slow, as soil is so heavily contaminated and has to be divided into two categories, high level and low level. Disposal of this soil must be carefully managed and deeply buried. There will be other unusual problems to deal with, such as the need to manage the new sarcophagus microclimate. As the structure will be so big it could even rain inside, so a process of using natural air current will have to be devised to keep the moisture levels down.

In its September 2005 report on Chernobyl, the International Atomic Energy Agency (IAEA) has done nothing to enhance our learning and knowledge about the scale of the tragedy, as it adds further confusion by trying to find logical and finite answers while missing the whole human and environmental trauma. Further, this report has added unwitting support for the governments of the affected region's policy, declaring the Chernobyl disaster officially over. The IAEA report adds legitimacy to the governments' policies of repopulation of previously evacuated areas and recultivation of lands within radioactive zones. The IAEA reinvention report on the consequences of the disaster will be used to support the building of a nuclear power station 25 miles from the exploded reactor on the territory of Belarus and further encourages the vested interests of some who want an optimistic outcome to the Chernobyl disaster.

The IAEA report should be greeted with some suspicion when you consider that an agreement signed in 1959 between the WHO and the IAEA hinders the WHO in its freedom to produce material regarding the consequences of Chernobyl without the agreement of the IAEA. The primary objective of the IAEA is the promotion of nuclear power plants in the world. Article III of the agreement states,

'The IAEA and the WHO recognise that they may find it necessary to apply certain limitations for the safeguarding of confidential information furnished to them.' It imposes a kind of '*daktat*' on the WHO, thus ensuring the absolute control of information on the dangers and risks of nuclear energy.

Despite the final closure of Chernobyl on 15 December 2000, the words 'the next Chernobyl will be Chernobyl itself' often run riot around my brain, haunting me with the image of a crumbling sarcophagus crashing down on a burnt-out reactor and releasing even more radiation than the original accident. Knowing that only 3 per cent of the original nuclear material was expelled in 1986, leaving behind 216 tons of uranium and plutonium still buried inside the exploded reactor, is a chilling reminder that the closure wasn't the end, but the beginning. Tall cranes still stand, forbidding, leaning over the shells of Reactors 5 and 6, frozen in time, relics of when time stood still on 26 April 1986.

Since the original disaster in 1986, over 400,000 people have been relocated from the devastated areas. They have become environmental refugees. The area they left is now a radiation desert composed of depopulated no-go areas covering many thousands of hectares, fenced off with barbed wire. It's a haunting monument to humanity's destructive capability. Evacuations will continue long into this century because new areas of contamination are found every day.

Those who remain in the zones have to change their lifestyle drastically. For instance, they are supposed to change their clothes twice a day, they may not walk for more than two hours a month in the contaminated woods and they may not grow their own food.

Imagine the life of every individual in the zone, who must organise his or her life around the radiation level charts printed daily in local newspapers. They live under the sword of Damocles. A mere glance at these charts will make it clear how complex and unprecedented the situation is. One in four of the population of Belarus has become an innocent victim of Chernobyl, of an atomic Armageddon, hostages to the hazardous aftermath of radiation. When the first radiation maps appeared in national newspapers, it immediately became apparent that the contamination was much more extensive than had previously been believed.

While travelling to Gomel on our first day in the radiation zone, we stopped to

as limiting the amount of time spent outdoors, keeping doors and windows closed and leaving rubber boots and farming shoes outside the door. There are rules for indoors as well, such as implementing a much stronger regime of cleanliness, washing living rooms more often, spraying ashes with water before

talk to some farmers to hear if Chernobyl had changed their lives in any way. We were unprepared for the stories that we were told. It was mainly the women who spoke. They cried bitter tears of anger and frustration at how their lives were locked into the effects of Chernobyl. Farmers are instructed to observe ludicrous rules, such

cleaning the fireplace and drying dishes, crockery and cutlery thoroughly in order to limit the doses of radiation. People in the zones are encouraged to have a more nutritious high-protein diet, which is supposed to improve resistance against internal radiation and to lower absorption of caesium-137 and strontium-90.

One woman spoke angrily about the 'death money' they are given by the authorities, which is supposed to be used to buy uncontaminated food. She pointed out that since most shops, particularly in the villages, are empty, people are forced to eat their own home-grown contaminated food.

Farmers are given itemised lists of vegetables, the most radioactive ones on the top of the list and graduating downward. One of the farmers translated the list and gave it to us as follows: cabbage, cucumbers, tomatoes, onions, garlic, potatoes, carrots, beetroot, radishes, peas, beans and sorrel. He told us that redcurrants and blackcurrants were the most dangerous fruits and to a lesser degree cultivated strawberries, apples and pears.

Farmers are warned about organic farming practices that can increase the levels of radionuclides. For example, they are supposed to check their fire ashes for radiation before using them on their compost heaps. They are also advised that after harvest, old leaves and other vegetation that would normally be composted should be destroyed.

Housewives in the zones told of even more bizarre instructions for the preparation of food. Vegetables and fruit were supposed to be washed in 'clean water', though of course, no one is told where this clean water is to be found. The removal of skins, the external leaves of cabbages and the heads of root vegetables, they are told,

will also reduce the radioactivity. All fruit and vegetables are to be washed at least five times.

Regarding the keeping and feeding of livestock, the farmers told us that for hay and pastures they had to try to select the least contaminated areas. Cattle should not be let out to graze when grass was less than 10 cm high, so that they would not touch the earth. It was recommended that cows and goats should be fed in their stalls and only by foodstuffs from the least contaminated areas.

Preparing milk for family consumption was another crazy story. The farmers were told that to reduce the radioactive contamination, the cream had to be separated and the fatless milk used for making cheese. The remaining liquid should not be used, but there were no instructions about how to dispose of it. Because radionuclides remain in buttermilk, the farmers were told that they must use an elaborate filtering system to try and clean out the radioactivity.

People were told that if they pickled their meat the radiation would be reduced. They were also supposed to wash the meat constantly with clean water. Once more, no one had any idea where to get the water or what to do with it once it had been used.

These men and women had few choices in their lives. As one woman said, 'If we don't eat the radioactive food, we will starve.'

Approximately 2,000 towns and villages have been evacuated. These depopulated areas will be uninhabitable for hundreds of years, even according to the most optimistic estimates. At least another 70,000 people in Belarus are still awaiting evacuation. But to where? Many of the evacuees that we spoke to have ended up being uprooted for nothing. They discover that the areas where they were moved to became as contaminated as the places they had left. The situation is complicated. In one village we discovered that the houses were fit to live in, but the fields around the village could not be used for growing food. In the next village the opposite was the case.

Farm workers from a collective farm situated not far from the radioactive area said their children complained about always being sick. They ate contaminated food and they drank contaminated milk. One mother said, 'Who cares about our children? No one.' Standing, staring, listening to her mother talk, a child of about 10 responded to my question about her health by saying, 'It's just the sickness of the children. That's all.' The farm where they live produces 8,000 litres of milk from 900 cows every day. Milk production continued during and after the accident. Despite the officially permitted levels being altered many times since 1986, the farmers have been told that their milk is within acceptable limits.

One woman said, 'We wanted to be evacuated. We've waited all this time. Why are we being left here? To die, is it? Why have they not fenced off this land like over there? Everyone comes with their machines. We hear the bleep-bleep-bleep and then those men go away and don't come back to help us. We all want to go.'

This woman's story is unfortunately just another example of the crisis of a poisoned land and a broken people. Professor Vasily Nesterenko has commented, 'During World War II, 280 of our villages were burnt to the ground by the Nazis, but now 3,000 villages are burning slowly from radiation poison. They should be evacuated, but our government refuses. Our death is slowly emerging.'

I stood recently in the village of Bartlomyvka, a small village in the heart of the contaminated zone, and witnessed this ancient village being demolished because of radiation levels. It was a terrible sight to

94

behold – all that life, all that history, all those neighbours gone and scattered to the four winds. I knew everyone in that village. I stood and watched the home of Vika and Dima being unceremoniously bulldozed into the ground. This solid timber house which had given shelter to her family for several generations was no more. Their home, along with dozens of others on that day, disappeared, wiped off the map into oblivion. I asked her how she felt. She replied, 'I have no words. It is locked in my heart and there are no words today. This is the war of all wars. Chernobyl means death.' Even those who do not cry sit like tablets of stone, their eyes full with unshed tears. They cry for the past, for what they've lost, for those who've died and for those who have yet to die.

With the benefit of hindsight, we can now state that when the evacuations from the exclusion zone finally took place, they were grossly inadequate and, for many, came too late. They were lengthy and disorderly and took 12 weeks, until June 1986, to complete. From then on, there was a bizarre belief that the fence surrounding the exclusion zones would somehow control the movement of radiation.

We visited two evacuee towns outside Minsk. In Shabani over 10,000 people lived in a concrete jungle of 15-storey, characterless, functional buildings. Malinovka was no different. It housed 11,000 evacuees, of whom 5,000 were children. Most of the people were peasant farmers who were now landless and jobless. Who wants the skills of a farmer (who may be illiterate) in a big, sprawling city? The stress and strain caused by the evacuations has resulted in huge social unrest. Six out of every 10 marriages now end in divorce. Many of the people we spoke to said that their living standards had worsened and that added stress had been created by the break with their cultural and historical values which had been caused by the forced removal from their lands.

In both evacuee centres we heard how the people, referred to as 'the Chernobyls', are often hated and isolated. This stigma has deeply hurt and offended them and causes terrible division within their new settlements.

The stories shared with us were difficult to hear. A young woman told us that at the time of the evacuations she was pregnant. She subsequently gave birth to a little girl called Lydia, who was born with stunted growth, immune system problems and a severe thyroid difficulty. The mother had been made sterile by the radiation doses she received. The father was now unemployed and tried his best to eke out a living as a labourer.

According to the director of the evacuee organisation in Malinovka, 95 per cent of the children are ill. They are tested only every six months and little information is given about the results.

Seven women evacuees came and told us their stories. Tanya told us how she had to bring her furniture with her, even though she knew it was contaminated, but she had no option because she had no

way from Moldavia. They had said, 'It's better to die slowly from radiation than to die from a bullet in the war.' Tamara wanted to take her family to a clean area for a short while after the accident, but she

money to buy new furniture for her otherwise bare flat.

Tamara spoke about the 'black sticky rain' that fell after the accident. It ran down her face and stained her clothes. She also revealed that when the local villagers pleaded with the authorities to evacuate the children and pregnant women, they were rebuffed. The authorities chose to evacuate 100,000 cattle in preference to the women and children. Tamara spoke about some of them, who had come all the

was told she could be dismissed from her job if she 'dared' to leave. She needed the income from her job, so she and her family stayed.

Tatiana told us how she was compelled to feed her family radioactive food. She had no other choice. Her husband had died the year before, aged 36. She was convinced it was because of Chernobyl. She and her family had lived in the zone for six years, awaiting evacuation. 'I remember the early days so well, people said that if the sparrows flew in a town that it meant it was safe. I was on a bus one day and the driver had to stop as birds kept crashing into the windscreen like they were crazed, blinded. It was like a kind of suicide. I will never forget that.'

Natasha cried as she told us how her husband had left her and their two children, aged five and seven, because he couldn't take the added pressure in their lives. Her son is extremely ill and Natasha has to carry him. Her added agony is that the authorities will not recognise her son as a 'Chernobyl invalid'. They weren't evacuated until 1993.

Galina and her family have to share a tiny apartment with another family. She receives $5 a month on which to survive. She never knew anything about the accident. Her children remained outdoors during the crucial days while the reactor burned. After her village had been 'de-contaminated', her son became seriously ill and never recovered. He now had a severe liver problem, but was given no medicine for it.

Marsha's son, who was born in April 1986, never left hospital for the first three years of his life. He appeared perfect at birth but almost immediately fell ill. Now he couldn't speak, and he had heart problems. Marsha's husband was a liquidator.

Anna talked about her daughter Tanya, who was also born in April 1986. Tanya was now severely affected by a variety of health problems. When she had been tested as a baby, the dosimeter had gone off the scale. Anna had 12 other children, all of whom had been diagnosed with thyroid abnormalities. One of her children had had her thyroid removed.

All seven women and their families were representative of the evacuees. Many of them were poorly dressed and unable to work because they had to take care of their sick children. Each of their children had a special government certificate declaring them 'Chernobyl invalids'. This stigmatised them as victims and was a permanent reminder of their plight. None were fit enough to go to school. They all wet their beds at night. Thanks to the courage of these women, a small community organisation had been formed to try to get help for the evacuees. We found very few such community groups in our travels. It's important that people should start to empower themselves. Most people were afraid to organise for fear of repercussions. The isolation and stress in the lives of these innocent women were unbearable.

One elderly woman refugee we visited showed us how, for the old people especially, the evacuations were the beginning of the end. Terrified of the city she was now forced to live in, with no way of sustaining herself financially, she was totally dependent on the charity of her neighbours and on the 'coffin money' of $3 a month. This was the first time I had heard the terrible words 'coffin money', a term used for the miserable contribution given by the state to evacuees. She had no shoes, just paper on her feet. This woman told us, with great sadness, of how she had once had a house, a pig, some chickens and land to grow vegetables. For the land that had supported her family for generations, and for all of her worldly possessions, she was given the sum of $5. This old woman told us she had nothing left to live for.

The same stories cropped up in every evacuee centre we visited. In an evacuee village outside the city of Gorky we were shocked by the living conditions. Houses of very poor quality, built in a hurry with the cheapest of materials, were inadequate substitutes for the lovely 200-year-old cottages the evacuees had left behind. When we visited this village in five feet of snow, the houses reeked of dampness and had little or no heating. Some of the people had no electricity, no medicine and little food. Children, lurking behind the doors, had scant clothing and no footwear. These families were trying to survive on $6 a month. One family had lost two babies and their third was seriously ill.

In the same evacuee village we sat and talked with the people about what they remembered about April 1986, and this was where we heard again about the 'black sticky rain'. It was very unusual and marked people's clothing and skin and took a long time to wear off. Enquiring about this strange rain, we were later told that it had been caused by the effects of the 'cloud seeding' that had been ordered by Moscow. Apparently several military aircraft were ordered to spray the radioactive cloud which hung over parts of Belarus and Russia in order to force a change in the weather. This calculation decided who would be contaminated and who would not.

The threat to the 'hidden ones' is very real. One time we were secretly being taken around a deserted village in the Gomel region, and we came upon an old man and his sister. They had been evacuated initially in 1987 but had subsequently returned because they were unable to settle into their strange new urban life. They had been caught many times by the police and each time had been forcibly removed, only to return again and again to their house. The old family home was cold, dark and squalid, but nonetheless home.

One disturbing aspect of the refugees' story is how contaminated houses are being robbed of their furniture and possessions and sold outside the zone; the houses are stripped and carefully disassembled, then are sold and rebuilt as holiday homes in places as far away as the Black Sea. The former villagers see this as a further violation of their homes and lives. There is also the serious risk of spreading contamination to the unsuspecting buyer.

I remember returning to the village of

Lipa, where we had done some extensive documentary filming years earlier. I was deeply affected by the changes in the intervening years. What had once been a beautiful, proud village, even in its desertion, had been ravaged by unscrupulous, nameless men. They came brazenly to the village. Numbers had been crudely daubed on each house. Each timber, each window, and each door had a number. The houses were disappearing. Whole streets were gone. It was a disquieting sight.

As we stood in the middle of the village, the silence was suddenly broken by the sound of sawing. I turned and saw a movement in someone's house. The three of us moved towards the sound. We stood at the door and surveyed what was being done. Two men were busy sawing up the floorboards. We asked them if they had the house owner's permission. 'No,' was the gruff answer. 'What will you do with the timber?' we enquired. 'It is already sold, and when we have taken this house apart we will rebuild it somewhere else,' they replied. 'Do you know this timber is radioactive?' we asked. They responded, 'We don't give a shit.' 'So who has given you permission to come in here and to take houses apart?' we asked. We didn't receive an answer. They just shrugged their shoulders and gave a look which told us they didn't welcome our interference. They were pretty rough-looking men, and we took a risk in asking them any questions, so we accepted the signal they

gave us and retreated quickly, wanting to know more but not daring to ask.

As we walked away, the sound of their sawing filled the silence. I felt that a terrible thing was being done. This village had been like a burial ground, and these men were violating and desecrating this holy place. We were to find the same picture in many deserted villages.

My mind kept going back to our first visit to the village of Lipa when filming *The Day of the Dead* and to the villagers we had met as they revisited their old homes for that special couple of hours to pay homage to their dead.

Getting permission to film there had been difficult, but finally we got to be accompanied by the special military force who work in the zones. We started filming the cars and busloads of evacuees as they presented their paperwork allowing them to return on this day to the village of their ancestors. The guard swung open the gate, a gate which carried a special radiation sign with the words in Cyrillic, 'STOP, do not enter. Forbidden radiation zone. No fishing, farming, grazing of cattle. No picking of mushrooms or berries.' The people entered the outskirts of their village under the strict control of the guards.

Finally our turn came. Under armed escort we entered the village. Silence descended. Everything and nothing had changed. As we drove down the main village street, we looked and saw the little

timber houses, with the traditional seat for the old people outside each house; the ornate drinking wells; a house with washing still on the line, too radioactive to be taken down; a child's bicycle up against the fence of a house, a doll sat in the window staring. An eerie feeling crept over me.

In the distance we saw a cluster of trees on a hill. Our guide told us that this was the traditional place in every village for the graveyard. It's on a height so that the dead can watch, listen to and be part of what happens to the village below. Droves of people covered the pathway up the hill, slowly winding their way to their ancient sacred burial grounds. We parked and prepared our equipment, conscious that we must be obvious to the people around the graveyard. But we were received warmly, as they knew the power of what our film could do in telling the outside world about their plight.

My job was to take Geiger counter readings every 10 minutes and to calculate how long we could safely work in this contaminated area. We knew we were on borrowed time, as we couldn't stay in this area for too long without running a very serious personal risk. I monitored for the following four hours and constantly added up the total dosage to ensure we didn't overexpose ourselves.

We stood and watched as the village people recongregated, surrounded by armed guards. They embraced each other,

cried with each other over the lives they had had to leave behind, and yet celebrated that they were still alive. It was the whole gamut of emotions. Families

The Day of the Dead

encircled their family graves. They laid clothes over the grave mound and shared a meal. At the head of the grave a dish with a little bread, a glass of vodka and some hard-boiled eggs was left to be symbolically shared with the spirit of the dead relatives. I envied them their strong connection with their dead, a connection that is almost lost in the Western 'developed' world. We discovered the importance of old people to all Belarusians. The extended family is

104

everything, and the old people are treated with enormous respect. Not to be able to visit the graves of their loved ones is a loss that cannot be quantified for Belarusians. It's as if a part of their heart is wrenched from them. Old people told us of the terror of being evacuated and the loneliness they now feel in being separated from their community.

I saw one old woman banging her head against the gravestone of her son, who died in 1986. She cried out her grief. Looking around the graves, I noticed how many of them dated from 1986. A man pointed to the land of his forefathers that he, too, had once farmed. He wished that some day he could return behind his horse and plough to bring life to these fields again – a wish that will never come true.

While the guards were preoccupied, one old woman, Tanya, secretly brought us to her home. We walked with her through the rooms, now left to rot in the house she had worked so hard to maintain. She pointed to the corner where her bed used to be, the corner where the kitchen dresser had stood for decades. Tanya talked about how she was forever 'rooted like a tree' to the earth there and how she was now 'withering and dying'. She looked around her little farmyard, shaking her head in disbelief. She bowed to her house, said goodbye to each of her trees. 'We didn't just lose our village, we lost life. Chernobyl is like a stone in my heart, always there, always heavy.' Her husband joined us and

added, 'Chernobyl means death. The song of our village has come to an end.'

We went on to film around the streets of Lipa and found a family hiding there. When we asked why they stayed there despite the levels of radiation, the man knelt on the ground, picked up some earth and, with tears running down his face, wetting the earth, he told us, 'This land is sacred, this is the earth of my ancestors – this earth is my soul; take me from the earth and you take my soul.'

We decided to enter one of the deserted houses to see what remained of people's lives that we could film.

As I stood in the middle of the road I couldn't decide which house to go into. I finally settled on a house with peeling blue paint, with a child's tricycle left propped up against the gate as if the child had just popped in for a moment and would be back any second. As I opened the garden gate and walked to the front door, I had an impulse to knock on the door or to cough out loud to let the people know there was someone outside. I was faced with a padlocked door. I felt like a thief as I 'broke and entered' the house by putting my boot to the door. Underneath my mask I became conscious of my own breathing, which appeared to come from someone or something else. A sense of unrest and uneasiness in the house surrounded and engulfed me.

As I walked through this little place that once was full of life, I imagined what it had been like on that fateful morning in 1986. People had been innocently sleeping in their beds or eating breakfast, when suddenly army tanks came rolling down their village streets and men dressed in strange suits shouted to the people to get out of their beds, to leave their homes, leave everything behind. What must that have been like for the people? I felt immobilised by the wave of emotion that came over me. I had a deep sense that I wasn't alone there, that the spirit of all those who had lived, loved and died there were somehow hovering. I remember standing in their bedroom, looking at the imprint of the body that had lain in that bed on that fateful morning. I remember seeing a little table laid for the breakfast that had never been eaten. It lay before me, frozen in time for me to witness all these years later. I looked around and saw photographs of the people, the grandparents, the parents, the children. The people had left, but their images lived on like their souls. An old Christmas card lay on the floor. A calendar hung on the wall, showing the date: 26 April 1986. The day the world changed forever.

I felt intrusive, voyeuristic even. I was the invader coming to see what the human species had done to themselves and their habitat.

Continuing to walk through the village, we came upon what had once been the village medical centre. Now all that was left were empty babies' cots and beds, empty medical cabinets, pieces of children's clothing, some small baby blankets and baby bottles. Adjoining the medical centre, we found what had once been a small library. Half-burned books were all that remained. What had once been a source of pleasure and enlightenment had become a harbourer of danger, as the pages absorbed and held the invisible radiation.

I thought about the people who had once lived there. All were peasant farming people who had been uprooted from their lives and land, people who had lived in harmony with nature since time began, now rootless and useless in an alien urban environment where there was no need for their farming skills and where they were perceived to be unwanted interlopers.

Another alarming development is the growing movement of refugees from war-torn parts of the former Soviet Union such as Azerbaijan, Kazakhstan and Uzbekistan. Fleeing from one horror to another, these unfortunate people are moving into the radioactive houses in deserted villages. When we visited a deserted town called Stari Vyskov, we found that the school had been reopened and was now educating children from 11 different former Soviet republics who had moved into the area because of war in their own countries. For the original inhabitants of these villages, seeing strangers move into their ancestral homes and lands adds to their heartache and loss.

In a startling development, I have recently been advised that new houses are being built in the most contaminated zone, known as the 'purple zone', and that people are being encouraged, with financial incentives, to return to live there. Local farmers are being directed to start ploughing again and are sowing crops in highly contaminated fields. President Lukashenko has called for the establishment of huge farms to grow peas, to raise cattle and to grow onions. Leaders of the opposition, along with foreign experts, all concur in their verdict that this directive from the president will end up with the food market being flooded with contaminated food.

A further sad footnote to this story is that UNESCO reports that poor people from throughout the three affected countries are moving back into the contaminated regions in order to receive special state benefits. In 2003 the state started to restrict the payment of the 'coffin money', leading to further impoverishment.

The power and strength of the victims' stories moves me deeply. They are by people who speak and write about a new way of counting time – 'before and after Chernobyl' and by children whose memories are divided into 'life and death'. 'Life' was before, 'death' is now. Their testimonies are a powerful witness to human suffering and endurance as well as acting as an ominous forewarning of what could befall any of us if we do not take heed and listen.

Svetlana Astsnova's story
When we start speaking about the disaster we should begin by speaking about our lives. We now consider our lives before and after Chernobyl, and I remember perfectly well the time when it came to us. It happened all in one night, but nobody knew it. It was very dangerous because the weather was unusually hot during those days. Usually in April the weather is rather cool but during those days it was very hot. All the people were very glad to be outside and to get a little bit of sunshine.

It was Saturday and Sunday and people were in the streets, but something had happened because in the evening of the next day my brother telephoned me and he told me something very strange. I didn't understand him very well. He was supervising some military department, a kind of service there in our atomic facility. He told me to keep inside and not to go out. He said something was strange,

because inside the institute it was quite okay. The dosimeters didn't show anything. But outside the dosimeters fell behind the scales. It was strange, they could not understand what was going on.

Later on, some hours later, they tried to get in contact with the special committees for our civil protection. They could not give a direct answer because they weren't allowed. They didn't know themselves, and only some days later we got to know what happened. Certainly it was too late, because our thyroids had already absorbed all the radioactive elements. Our thyroids were ruined.

On 1 May we had our usual May Day parades and all our people were brought to them from all over the country. They were not cancelled; it was very strange. People knew little about the accident but, you know, when danger is not seen, it is not felt. It was very unreal for people to understand that they had already been contaminated. When I met my pupil in the street, she was walking with her baby in the pram. I told her, 'Go away, stay at home.' She said, 'Why should I? What's the problem? The weather is so nice.' They could not understand, they didn't believe; it was like a silent death. I remember I was shocked when we walked past our parade stand, with all the leaders and the army, to see they were dressed in white radiation suits! We were there with bare arms and legs! What was the matter?

I am a teacher and I was asked to

108

volunteer to take care of these children called 'Chernobyl Children'. They were the first to be brought to a special place outside Minsk. I had a chance, a tragic one, to work with very little children. They were five, six, seven years of age only and they stayed for two months in Minsk Hospital. They were very ill. They were deprived of their mothers for the first time. They were all country peasant children. Some of them were very self-restrained – they were very shy. They couldn't get accustomed to this new environment. Those children were taken away from their homes all of a sudden and now they were very ill. They were frightened and confused.

It was the first time I felt the reality. The most tragic thing was when the children were arriving in dozens of buses. We were put in some special suits for protection from the children. That was such a dreadful sight. I took off my suit to comfort the children. My job was to decontaminate them. I had to take off all their clothes and wash down the children. We didn't know what to do with those clothes. They were very contaminated. All their clothes were taken away. But then we had the problem of where to take those clothes. They could not be burnt, because all the radiation would go up. So we decided to bury them. Again, where to bury them?

But the most tragic thing was that the children themselves were radioactive.

They radiated to such a degree that when we were checked later on, I myself started to radiate – my liver, my thyroid and especially my feet. Their clothes gave off such a great radiation that I was afraid. When I looked at the dosimeter it began to move before my eyes and make a noise – d-d-d-d-d-d – oh, it was really startling. I was scared very much indeed myself.

We applied to our factories and authorities to help us because those children came from the most contaminated zone and we wanted to have them taken somewhere else which was safe. We received little help. They were practically vanishing before our eyes. They were becoming pale, even blue in their faces. Their blood pressure was terrible, they were vomiting all the time. My little ones could not stay in bed because they were crying for their mothers all the time. They struggled with me, sometimes they threw my glasses off. Something terrible was happening to all our children. They could not eat. The state budget was so very poor, and we got no help.

It was like a nightmare. Can you imagine? Some children were given special identity labels – red, green or yellow. If a child got a red label nobody could approach him or touch him. If he had a green label, that was bad but not so terrible. Yellow was okay.

Another thing – we teachers that volunteered had to bring our own children, since we had no one to mind them. We tried to keep them aside somehow; we put them in a special building and they were separated, but still they are children, you know, and we could not separate them forever. They played together, and that's why our children got some portion of the radiation. That was the most terrible thing. I have only my one son, you see.

These children were blocked on one of the floors. They were not allowed to go anywhere. That depressed them so much. Can you imagine? When parents came to visit their children they were not allowed to see them! They didn't know what to do. Some of them were so desperate that they took their own lives, believing they would never see their children again. This lasted for 60 days.

I will never forget the mornings. It was terrible. We had to line them up in the square, and it was to become the routine of our camp life. One by one the children fell. They could not stand even for five minutes. They fell down and we took them away in pairs sometimes. It was a dreadful sight. I then began to realise that something so awful had happened.

Svetlana Polganowskaya's story
In Chechersk, on the day of the accident, there was an open air concert, so all the people were in one place. Coming home it got very dark and started to rain very heavily. My children were playing in the sand with other kids. They were just in front of us but we could barely see them. Everyone got very scared, especially the children.

The rain was black, the sky and the ground looked the same colour. It was like sand falling down on us. When we got back to our apartment I decided to give the children a wash in the shower. I was taking off their clothes, I could feel something unpleasant on their skin – it was not sand – I can't really explain. Then in the bath I heard some noise – like metal pellets falling into the bath. We had no idea what this metal grit was.

The May celebrations that year were like they always were, except this year the smallest children took part. The girls had ribbons in their hair and the boys wore shorts in the hot sun of early summer. They stayed in the town square all day as the music played and the military paraded. I watched on as a proud teacher and mother. I had no idea what danger they were in. We know now that it was on those first few days that many of the children developed cancer.

We first noticed that there was something wrong when the children of the party officials were being taken out of the school and sent to places like Minsk. Stories began to filter through from down in the south, like Narovleye and Bragin, that women had been forced to have abortions. No one was telling us anything. There was nothing on the radio or the television and the telephones weren't

Svetlana Polganowskaya

working. When Gorbachev came on TV he said everything was fine, under control. But I knew it was not the truth. I noticed strange yellow-green colours in puddles of water and when I asked the authorities about it they told me, 'Be careful now, you are too smart for your own good, keep your mouth shut.'

Then the officials came and told us that because of the accident that all the children would have to be taken someplace safe. I was in panic, so were all the other parents – we didn't know what it meant. I was so upset because my three children were leaving to go to the sanatorium. My father and I were standing near the buses seeing them off. My father remembered the war and he compared the situation to when the war started, when all the children and their mothers were evacuated. They went without their parents to different parts of the Soviet Union and even now, 30 to 40 years after the war, some of these parents are still trying to find their children.

I organised a protest march in Moscow and uncovered real evidence that the local district leaders had lied to Moscow by telling them that all our regions were completely cleared of human life and that all of us had been evacuated to new houses. But we were still living there with our children! When I spoke out about it they punished me by saying that none of my family would ever be evacuated and that we would rot in Chechersk. They tried to shut me up by arresting me many times. They even handcuffed me on television! They fear no one.

I am so tired now and inside I am empty. This has been going on for 20 years and it hasn't got any better. We have just become used to it. No one has been honest with us, no one has told us what to do. We have been left to fend for ourselves.

I have seen and heard too much for this lifetime. I wanted to know more so that we could understand. I went deeper into the zones to find out things. I brought my young daughter Lelia with me. I didn't know then how dangerous this was for her. She now has many problems with her thyroid gland. Down in those places we saw the deactivation of houses. They were washing them down. I saw them remove soil from the gardens. I saw a gigantic pit where they buried herds of cows and other animals.

So much has been destroyed. So many lives touched and damaged that it's hard to see into the future. I was happy once. All of my children came from love.

In some ways I think I am lucky because I understand the danger of Chernobyl, but then some days I wish I didn't know anything; it would be easier.

But it isn't for me that I am sad. It's for the youngest people, the children, who have dreams and hopes. I wish I could say to my children that one day everything will be okay and things will get better, but I know in my heart I can't say that. My young son looks at me with hope and innocence and there is nothing I can say. I have to feed him with what I know is dangerous food, I feel guilty but I have no choice.

We can only live for now and as each day passes I know that things can never get better. Radiation has no smell but we're changing, our blood, we have genetic markers, the land, it's all changing. Chernobyl is in my house, in my children. With the shadow of death you understand everything. Cherchesk is a place with no future – for me it like a long winter with no spring.

Liquidator families in Gomel

Natasha speaks: 'My husband was beautiful. He was a great sportsman, really healthy, never ill. He volunteered to work as a "liquidator" and left us in mid-April 1986. He never returned. I saw my lovely husband for the last time, standing at our door, hugging our two children, smiling. They tell me he died in 1991 of a massive heart attack at the age of 34. I don't know the truth. He never phoned from that place because of some secrecy. Some of the men told me afterwards that he had been making special dumps for things that had got radiation, like buses, tractors, bulldozers and such things. I think he got poisoned from that place. What are we to do now? No father, no husband.'

Nina speaks: 'My husband was a militiaman and was conscripted to a "top secret" job on 28 April 1986. He stayed away for one month. He returned and has never been the same. He is unable to speak about what happened during that month. He has fallen silent, as if dumb. He is in hospital most of the time. He has problems with his liver, pains in every joint. He is a changed man, always tired and distant. He never talks to me or the children any more. We are like total strangers now. I don't know what to do. The government does nothing to help us. We just get $20 a month compensation. Life is hell.'

Anna and her mother-in-law sat and talked to us about their life in the evacuee village of Uza. Their eyes immediately filled up with tears. Anna, wringing her handkerchief in her hands, recalled her Chernobyl story: 'My village was Slobada in the Bragin region. I first heard about Chernobyl on the "Voice of America". It was an alien word to me but one that would haunt me forever. I was frightened. I had two children, aged six and nine. The militia came first with buses full of people from another village 10 km from Chernobyl. We had to take in these people. We felt lucky that our village was safe. But some days later they came with more buses and took us all to another village 30 km from Chernobyl and forced us to leave our village. Our village had become poisoned too. They told us they would make it better and we would return three days later. But we never returned.

'My two children were taken away for something they called decontamination. I did not see them for four months. My God, can you imagine? I cry now remembering. It was terrible. My children were stolen from me. We had no news of them during that time. The separation, the not knowing, was something I can't describe to you. Then, when we were forced to leave on the buses to come to this new place – you don't know what that was like for us. For the old people. They couldn't understand what was happening. Many of the old people had to be forced out of their houses.

'My God, you should just see that beautiful place we left. In my dreams every night I return with my family to our home. The river, the trees, our garden, our house and land. This will never be my home. I cried for two years when I came here first. I felt like a dead person during those two years. Nothing is stable any more. Our place is replaced with just pipes and concrete here. Our beautiful nature. My children drank that poisoned water and they drank the milk; nobody told us not to. What have they done to us?

'My children are all sick now. The people don't want us here. We are like intruders to them. It's hard to get work because people call us "the Chernobyls".

'My mother and father couldn't live here, so they returned. Many of the old people return. They prefer to die of radiation than to die of a broken heart. We are allowed to visit them twice a week. Everything is checked for radiation by the guards when we leave the exclusion zone. It's like going to prison, but the prison for me is on the outside.

'On the "Day of the Dead" they allow us home for a couple of hours. We don't see each other for the tears. It's terrible. Our houses are all looted. They have even ripped out our floorboards and sold them. All our things are gone. We have lost everything. Can you imagine – they even took the photograph of Mama. Why? I don't know anything any more.'

Anna's mother-in-law told us, 'When I return on the "Day of the Dead", I have no feelings, just tears. I can't put all the tears into words. It was my birthplace. Our village was everything – the forest, the river, the sun. I don't recognise anyone any more. There are no single memories, just all these years flashing by my eyes. Chernobyl means death.'

Galina Rodich, 10th form pupil in Chechersk secondary school No 3

It was a sunny spring day. My mother and I went to the forest to get some saplings. We came back carrying little birch and rowan trees to plant them under the window. We were not far from the house when a storm suddenly broke out. The sun and the sky were no longer visible. Dust was getting into our mouths and eyes. We ran into the house. The storm suddenly subsided as quickly as it had started.

We were happy as ever, we enjoyed our
    lives,
Every green blade, every daisy that
    thrives.
To the nearby field we, carefree,
    walked,
About our lifespan with a cuckoo we
    talked.

Mummy's eyes have lost their light,
And her smile is no longer bright.
There is grey upon her head,
In her eyes we see the dread.

Chernobyl has brought us all woe and
    death
I'm young, but may soon draw my last
    breath.
Chernobyl, you mean nothing but
    misery and woe,
To millions and millions you dealt a
    death blow.

Nobody was aware then that it was Chernobyl's hot breath. Its black wing had covered Belarus. Radioactive dust had settled upon the forests and copses, the smart clover meadows and fields, the precipitous banks of the Sozh and Dnieper, the blooming gardens, and the roofs of houses. We didn't know that Chernobyl had burst into our lives like a black tornado. We were surrounded by invisible radiation and our thyroid glands were greedily absorbing iodine. We hadn't been forewarned.

When the sweeping Chernobyl catastrophe fell upon Chechersk and other villages, none of the residents had any idea about the disaster. As they listened to the news on the radio, they sympathised with Kiev residents, sighing concernedly about the destiny of the Polesye, and crying as they met the buses carrying the people evacuated from Narovlya. Their grief wasn't our grief. At the beginning of June the schoolchildren of Chechersk started being sent to places of safety. Mothers and grandmothers who saw us off couldn't help crying. Those children who were parting from their mothers for the first time in their lives were also crying. The convoy of buses got under way. My mother who was a nursery school teacher began to comfort the children. It worked. We were singing songs as we arrived at the railway station. The weather was hot. After two hours a diesel train took us on further. We went on singing. Mother joined in the chorus, but there was unbearable pain in her eyes. One girl named Marina asked her, 'Galina Alexeevna, why are the old women blessing our train with crosses?' Mum went down to the end of the carriage and burst out sobbing.

Today, caesium is lurking in the fields and meadows, and the green oak groves. Actually, the apple trees, the lawn and the rivers have become our enemies. There is caesium and strontium everywhere. It is eight years since the Chernobyl catastrophe happened. We have been to the forest just once. Picking strawberries has been ruled out and any kind of mushroom has become inedible. We don't pick bluebells or daisies to make posies, and we don't use them to make wreaths either. We must not do these things, we mustn't we mustn't we mustn't.

My daughter, my dearest sonny
Stop touching the grass with your
    hand
Don't walk when it rains, keep from
    swimming
Try not to sit on the sand.

My daughter, be careful, darling
Keep away from the bright sunny glade
Don't pick any flowers, ever
Do try misfortune to evade.

Do not! Do not! We always hear
Strontium and caesium. Take them
    away!
To walk upon the grass without fear
We want to freely play.

Two years ago my grandfather's friend sent us a letter. He saw a village being buried. It struck him, he stood as if petrified and couldn't walk away. He stood for a long time until no more than just barrows remained from the former houses. He started walking away, dragging his feet and bent low under the invisible burden of grief.

Every word of the letter was steeped in pain. Three days after the burial of the village, grandfather's friend passed away. His heart failed. Rare is the person who can bear a thing like that.

I was absolutely shaken up by that death and wanted to see the houses being buried with my own eyes. It is a queer sort of burial. A frightful scene. A hole is excavated next to the house. The firemen drive up and wash the house all over from the gables to the foundations with their fire-hoses to prevent clouds of radioactive dust from rising up. Then a huge iron monster on powerful caterpillar tracks pushes the house, overturning it into the hole, and that is the end of it. Yellow barrows remain. My heart tightened up and there was a lump in my throat.

Chernobyl is our pain. Each person's pain, everybody's pain. It is the pain of migrants who feel like refugees or people who have lost their goods and chattels in a fire. Only a bitter dream can take them back to their own houses, to their native villages, only a dream. Just once a year, on bright and sorrowful Remembrance Day, they will be able to visit the graves where their kith and kin are resting in peace.

On that April day thousands of Chernobyl 'refugees' lost the most important and sacred thing in their lives – their roots.

Chechersk has suffered this merciless and terrible fate. It is in the contaminated Chernobyl zone.

I wish to enjoy my life, be happy
To watch the dawns and sunsets all
    aflame
Why must I suffer, why? Do tell me,
Am I for anything to blame.

Our only fault is that we stayed on our native land which is thickly sprinkled with radionuclides and we have wilfully taken part of other people's sins for irresponsibility upon ourselves. The result was one of the most appalling catastrophes of the century.

Our Motherland is our Motherland, even if it is strewn with the ashes of Chernobyl. The people cherish their dear land, sick though it is. It has become twice as dear to many of them since the catastrophe.

For all these years we have been bearing the Chernobyl cross. Chernobyl isn't in the past, it is our present, our tomorrow. It's our future.

The other day we saw storks. They have come again to our suffering land and made their nests here till autumn. They are like people whose hearts are attached to a plot of their native land and never leave it.

What is it? Folly? Stubbornness? Just love? An inability to betray? Faithfulness? Life is going on in the area round Chernobyl. It marches on despite all the misfortuncs and diseases. The earth is alive. Give it a helping hand and don't take away what it needs so much, or what is simply necessary for it to live. Think of the people who inhabit the ailing earth.

Marina Migdasyova, Zhodino

Chernobyl… Many people can't understand why this word makes one feel like crying. For a long time I didn't realise why either.

People are feeling the effects of the explosion at the Chernobyl nuclear power station even today. There are various kinds of disorders and diseases. I was also not spared by the disaster, and was taken ill. Lymphoma was diagnosed. My situation was critical and serious, nearly fatal. There were many patients like me in the clinic I was taken to, and all of them were struggling for survival. Anyone who hasn't been through anything like that will never realise what life means.

In my 16 years of life I have understood one thing – we have grown callous. We often see a broken tree, a destroyed anthill or a nest, but only a few of us will stop and not just pass by, leaving it unheeded. This is very much like our own lives – some people have been shattered by illnesses, while others have been scattered over Belarus.

Lots of material has been published about the way of life in the danger area. The picture of those parts is appalling as you try to imagine them – abandoned houses, kitchen gardens overgrown with weeds, and stray domestic animals. The doors of deserted dwellings are squeaking, rending your heart. A slight breeze is stirring the leaves on the tree branches. There is not a single human voice. Silence. A horrible silence hangs over such villages…

It is awful. And yet for many people this has become everyday life. Everyone holds dear the plot of land where they were born. Many people will never see it again.

Chernobyl has brought people misfortune, unhappiness, and tears.

Until recently I could only lie and look at the patches of sunlight on my bed, thinking about my own destiny, and the lives of other people. All of them want to live. I have come to feel it particularly acutely during the past year. Death is all around us – a dry leaf that drops from the tree, a moth that dies in autumn. Maybe death brings long-awaited peace and rest for some. Unbearable pain sometimes makes sick people throw themselves under trains.

My ward-mates say, 'We are all like sisters, even more so sometimes because our relatives have turned away from us. As for us, we are in for a course of treatment here, two weeks at home and then back here again.' The women smile mirthlessly. There are so many of them here, both old and young! They hold many things back from me and sympathise with me, feeling that the less I know, the better. But I understand everything – why the hospital nurses in the treatment room are wearing rubber gloves, why many women wear headscarves though it is summer now, and many other things.

In the evening I could lie and listen to the rustle of trees. I saw swallows and swifts rushing by, quick as lightning. All living creatures were rejoicing at the summer sun and warmth. Only I was motionless on the sickbed, lying with my eyes shut and wiping away tears, which were coursing down my cheeks and dropping onto the blanket leaving large wet spots. I wanted to live and I was confident I would be able to survive. I paid no attention to the time, those six months when I stayed indoors, and didn't talk to anyone except my own family. I never tell anyone about my misfortune, or pain that has suddenly broken into my life. Never. No one. I don't want any sympathy, which often seems to be false. I'll do without any help for there is no other way around it. Let everyone think that I am still the same daredevil girl I used to be. I don't want any changes.

I have many things to do before evening comes – first school, next home chores, then I knit or do some reading. Only at night when all have fallen asleep can I talk to myself and support and comfort myself a little. Mine is a kind family, I love them and don't want my relatives to be unhappy. I try to be cheerful and make jokes, but sometimes it doesn't work. I'm optimistic about my future. As soon as I get into bed, memories come upon me. They are not always happy.

I fall asleep with the belief that tomorrow will be better than today. In my dreams I see my happy childhood where there is nothing to disturb me, where I can wave my hands easily, fly up into the sky like a bird and see our Belarus in the whole of its splendour from above. Belarus – the land I love and will love forever.

During Christmas week of 1995, Ali Hewson and I were visiting Belarus to deliver Christmas presents to the children of various orphanages around the country. Our last port of call was to our favourite children's home, the 'No 1'. We always like to keep the best place until last because it gladdens our hearts to see the power of intervention and its impact in improving the lives of the children. We had 'adopted' the 'No 1' and made it one of our model places where we managed, with the help of the Chief Doctor, Tamara, to raise the standards of care and make the home a loving place.

On our arrival we asked to visit each of the children's units. Nothing prepared us for what we were to meet on our arrival into the unit housing the smallest babies. About 15 cots filled the room, laid out head to toe. Dr Tamara motioned us to one cot just right opposite the door. Her normally smiling face changed and was now filled with deep anguish and sadness. I understood why when my eyes followed her outstretched hand, pointed at a baby boy. I glanced downward and saw that half the side of the baby's face was blighted with a very large tumour-like growth. In the place where the left eye and left cheekbone should have been there was this enormous

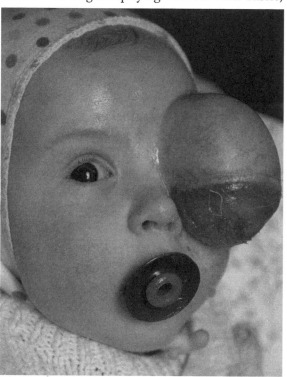

thing, oozing liquid. The baby couldn't move as his arms had to be pinned down to keep him from touching the growth, which was raw and foul smelling. I hate to admit it, but I almost got physically sick when I saw him. My initial reaction was to move away. As I passed from cot to cot, talking and playing with the other babies, I tried to forget what I had just seen. But I couldn't, no matter what way I turned. Even when I had my back to this baby's cot I could sense that he was following me with his good eye. In the corner of my eye I could see that he was making a great effort to turn his head and follow my movements. It was if he was drawing me

ALEXEI
SMARLOVSKY-
BARRETT

120

back, somehow begging me to look again, to ask about him, to reach out and touch him, to love him. I went over to him, stretched out my hand and stroked his face. His good eye latched onto mine and in that instant, it seemed that I looked into the soul of this little baby and saw his great courage and will to live. In those moments that followed I knew he was asking for something – he was asking us to save his life, as simple as that. I don't know exactly what happened in this brief encounter, but I knew we had to try and do something.

Dr Tamara told us that the boy was called Alexei, that he was five months old and had been given up at birth. His mother had been told about his deformity when she was eight months pregnant. She was advised to abort him but she refused, despite the fact that they told her he was brain dead and would definitely die. Dr Tamara told us that nothing medical could or would be done in Belarus and we talked with her about the possibility of getting him to Ireland for treatment. Ali took some photographs and Dr Tamara gave us an X-ray of Alexei's head to show to Irish doctors. We left Belarus shortly after and talked about nothing else on the journey home other than where we would start trying to get him admitted in Ireland. When we landed at Shannon Airport we came back into what felt like an unreal world. We had been so immersed in the devastation of Chernobyl for the past two

weeks that we were totally unprepared for what felt like the opulence of the festive season. The airport was full of Santa caverns, Christmas trees groaning with decorations and lights. Our ears awash with piped music of 'Jingle Bells' and 'White Christmas', we felt as if we'd entered an alien world. As we sat in the coffee shop we planned what we would do. Ali would race back to Dublin, making calls along the way to various medical contacts and see if we could get a hospital to take Alexei. En route she stopped and got the film developed, which meant she now had the possibility of making a stronger appeal, as we both felt that as soon as any medic saw little Alexei's plight they would be moved to take action. And we were right. Mr Michael Earley, the prominent Irish plastic surgeon in Temple Street Children's Hospital, agreed that if we brought Alexei to him that he would make a decision on what could possibly be done.

As it was Christmas week, everyone was full of the joys of the season and starting to wind down for the holiday period, but somehow we managed to persuade our Departments of Justice and Foreign Affairs to issue an emergency visa for Alexei to come in for urgent medical assessment. I found myself back on a plane on 4 January with an RTÉ two-man film crew, Kieran Fitzgerald (presenter/ producer) and Sam Gracey (cameraman), who were to cover Alexei's story for the news in Ireland. Coincidentally (or maybe not!), it happened to be on the eve of the Russian Orthodox Christmas, and at 4.30 a.m. on Christmas morning the three of us set out to collect Alexei at the 'No 1' home. He was sound asleep and was none too happy to be given such an early morning wake-up call. Right there and then I got a crash course on being an 'instant' mother, as he needed a nappy change and feed. Luckily a friend had given me a baby bag full of all the essentials. Not being a mother myself meant that I hadn't thought of practical things like nappies, a bottle, a change of clothes or, of course, a warm jumpsuit. We hurriedly dressed the little fellow as he watched me intently with his blue eye, the eye that was full of question and intelligence, but also full of anxiety. We bundled him up and took him with us to pick up the accompanying nurse, Irina, and our translator, Luda. They were waiting on the side of the road, freezing cold, as it was minus 20 degrees.

It was only when we were on the flight, when I could breathe a sigh of relief, that I looked properly at Alexei. I was shocked to see that the tumour-like growth had grown much larger in the intervening two weeks. The nurses had tried to cover it with a bandage, but it had quickly become soaked with a mixture of blood and pus. He was very unsettled on the flight; it seemed like he was in pain. I thought we would never arrive at Shannon. One of the things that distracted Alexei on the four-hour flight was his hands. This was a whole new experience for him, as they had always remained firmly pinned to his sides. His little eye was intrigued with the discovery of moving hands and fingers. On our arrival at Shannon, some of the staff came forward with baby clothes and food. There were photographers there too and I can still remember one photo which appeared, capturing my fear and Alexei's pain. It was shot just a moment after Alexei's newly freed hands had briefly touched the growth, which started a nasty bleed. The blood poured everywhere, into his mouth and down the side of his face and neck. The poor little baby was frightened, but not half as frightened as I felt as pure panic set in.

We eventually hired a car and the three of us headed off to Temple Street Hospital in Dublin, 134 miles away. It was a fairly testing journey for all of us, as we had no idea how ill Alexei was and we were constantly worrying that the tumour would start to bleed again.

In the interim period between when we found Alexei and when he arrived in Ireland, I had the job of trying to find a host 'mum and dad' as he would need people to come and visit him during his hospital stay and someplace to stay during his recovery time – that is, if he recovered. I couldn't find anyone who would take the risk of taking such a small baby that could die at any moment, and in desperation I asked the one person I knew who wouldn't refuse me, my sister Len. She and her husband, Chris, who have no family of their own, discussed the matter and, despite their anxieties, agreed to look after him temporarily. I gave them a photo of Alexei and they loved him from the moment they saw his little face. When I arrived into the foyer of Temple Street I was met by the two anxious faces of Len and Chris. They looked at Alexei snuggled up and sleeping in the carry cot I held and they both wept.

Alexei moved in with Len and Chris soon afterwards, as the hospital said that he was underweight and suffering from the effects of institutional care, and should be given loads of one-on-one love which would help build him up for what would be a very complicated operation. During these weeks Len and Chris showered Alexei with love and affection and he quickly responded. What had once been a sad and serious face became one wreathed in smiles of contentment. He also started to eat really well and very soon he started to put on weight. The problem that arose as a result of his new-found confidence and energy was that his hands constantly touched the growth, which bled frequently. Len had to start tying his hands to the bars of his cot each night. This broke their hearts, as Alexei cried and resisted the restriction, which I'm sure reminded him of his early experience. We started to call him the 'Incredible Hulk' soon afterwards

because as he got stronger he was sometimes able to free his arms, unfortunately often with detrimental results.

The date was set for the operation and Mr Earley, along with a highly specialised team who would perform the operation, called for a photo shoot with Alexei. At this stage Alexei had become the darling of the Irish public, as they had seen both the RTÉ footage of him and photos. A few days before the photo call we noticed that Alexei was very unsettled. He became very agitated and upset all the time. At the same time, we noticed that the growth had got not only bigger, but a lot angrier than it had been and it seemed like his face around the growth was slightly swollen. On the morning of the photo shoot, Chris and I got up early to wash and dress Alexei, as Len had gone to school. After his bath we laid him on the sitting room floor to dress him. I gave him a rattler to distract him while Chris put on the nappy. In a flash, the rattle touched off the angry tumour and it bled profusely. Within milliseconds blood poured into his good eye, up his nostrils and into his mouth. Alexei froze, then he gagged and gasped for breath. He was choking on his own blood before our eyes. I panicked and ran out the door, where I watched from a safe distance. Chris put his own fear aside while he frantically cleared Alexei's airways and nostrils. It was something special to see. Thanks to Chris's quick intervention, he managed to keep Alexei from choking. Very shaken, and with Alexei only half dressed, we drove to the hospital to meet Mr Earley, the team and Ali. The moment he saw Alexei he advised us that the growth had gone critical and that he would have to be admitted to hospital immediately. He said it was thanks to Chris that Alexei was still alive.

The operation lasted several hours and we waited to hear whether Alexei had pulled through. None of us knew if he would survive the trauma. We had been advised that Alexei would be in intensive care for some days and wouldn't regain consciousness for at least two days. But the spirit of this brave boy determined otherwise, and he was awake, screaming for food, within 24 hours. Ali and I went to see him in the special unit soon afterwards, and while we were upset at seeing his very swollen face and obvious stitches across his entire head, we were thrilled to see the tumour removed. This boy was a fighter and he went from strength to strength from that day onwards.

Len had received special leave and between her, Chris and a devoted team of friends, Alexei had someone to hold his hand all the time he spent in hospital. Thanks to that love and to the love he got from all the staff, he battled back to life and left the hospital to move in with Len and Chris for what was a very uncertain future.

We had never considered what would happen after the operation because we were afraid we would be premature with our plans in case he didn't survive. So we had a dilemma. What would we do now that he had survived? Would we get him fit and well to send him back to a mental asylum? Or did Alexei hold a new challenge for us to face? We decided it was the latter and looked at the possibility of adoption. Len and Chris were so bonded with him that they started the process of applying for adoption clearance. The problem now was that there was no adoption agreement in place between Ireland and Belarus, so what were we to do? In keeping with my belief that behind every problem hides an opportunity, I started to find out how to go about getting an agreement between our two countries. Not an easy task! Over the following three years we slowly travelled the quiet diplomatic route. There were times when we almost gave up, but thanks to civil servants like Helen Faughnan, our own Adoption Board and the Adoption Centre of Belarus, three years later we found ourselves with a new agreement brokered between our countries allowing Belarusian children to be adopted in Ireland, an agreement, I'm proud to say, that we played a pivotal part in negotiating. The test of this agreement came in October 1998 when Len, Chris and Alexei joined Ali, Helen Faughan and myself in Minsk. Before the court hearing could go ahead regarding Alexei's future, he had to undergo a battery of medical and psycho-

Alexei meeting President McAleese

logical tests in a variety of different hospitals. It was nerve-wracking because we had no idea what exactly they were looking for and what bearing it would have on his right to be adopted. In the middle of it all, Len and Chris had to get married again in a Belarusian civil ceremony, and then, finally, came the court hearing. I was allowed to join Len, Chris and Alexei in the courtroom along with our dear friend, former Ambassador to the United Kingdom and Ireland, Vladimir Schasny, for what was to be an emotion-packed meeting with Belarusian officials who would subsequently vote one way or the other on the adoption. It was very tense. You could feel the nerves bouncing off Len and Chris. Alexei was the only one oblivious to it all as he dangled his legs and chomped his way through a packet of sweets. The room consisted of a very large board table surrounded by about 20 officials. There was a judge, a dock and then three rows of chairs for observers. Evidence was called for and expert witnesses stood in the dock. We couldn't understand the language, which added to our fears. Then the time came for Len and Chris to give evidence, to be followed by cross-examination. My heart was thumping as they both stood, unsure of what would follow and yet knowing that these moments would define their future. The first question to both of them was 'Why would you want to adopt a child who has obvious physical and possibly

mental problems?' Their answers made my heart sing with pride and joy as they spoke about the love they felt for Alexei and the love they experienced back from him in return. They gave an account of their life together as a family. They recounted times like when he took his first step, when his first tooth appeared, when he said his first faltering words – all the simple but beautiful moments they'd had with him. Chris spoke about his love of Russian history and the empathy he had for the pain of the Belarusian people, particularly during the Second World War. He drew parallels between Irish and Belarusian history, a history full of famine and colonisation yet full of rich culture, literature and humour. He spoke about our common humanity and how, if they agreed to Alexei's adoption, that Belarus would not be losing a son but sending an ambassador for their story to the world.

When Chris finished I looked around the table to see if any of the faces showed how they would vote. I noticed that all of the women were crying. Very quickly afterwards some minor questions were asked and answered, then in a final speech the judge said the magic word we understood – 'da' ('yes')! I brought little Alexei around the table to greet and shake hands with everyone there who had decided his fate. He practically ran around the table, his little blond head bobbing up and down, smiling all the way. When we burst out of the room all our friends were waiting and I felt like we were wrapped up in one enormous hug.

Later that day we went to visit the 'No 1' Home and Dr Tamara, as she was his first mother and the one who had pushed us to bring him to Ireland for treatment. She wept when she saw Alexei, and the relief and joy leapt out towards us. Len asked her if we could have the name, address and phone number of Alexei's birth parents, as she and Chris would like to meet them. There were some anxious discussions between the Belarusians about this request, as they don't approve of the natural parents having any involvement in adoptions. But when Tamara saw how important it was for Len and Chris, she agreed and there and then we phoned Alexei's mother. She was very surprised to hear from us but was willing to come to our apartment to meet us later that evening.

It was decided that it was best that Alexei would not come to this meeting, as the expert advice given to us in Ireland was that he was still vulnerable and too young to understand the concept of two sets of parents. We also felt that Alexei would come to meet his Belarusian parents when he was ready himself.

Len and Chris were a bundle of nerves as they waited for the arrival of Alexei's mother, his aunt and cousin. What would they be like? What would they say? Nobody knew what to expect. The added pressure of having the whole event filmed

for a documentary didn't help to alleviate their growing dread of the meeting. Finally they came. Irina, his mother, was so beautiful and she immediately embraced Len and Chris. The meeting was powerful, almost indescribable, as they talked about her son, the son she was never allowed to hold, the son she had never seen. She asked many questions about him and told us that she wanted him to know that he was loved into life, and she and her husband would always love him, but that for them the ultimate act of their love was to give him away so that he had some hope to stay alive. As she said these words we all started to cry at the sheer courage and strength of this woman and the love she held for Alexei. She told us that because both she and her husband had been living as teenagers in the radiation zone, all of her pregnancies had been damaged. She was now in a special scientific control group where she is constantly monitored. Her dream was to have a healthy child. When all the tears were shed, all the hugging done, Irina and her family left on the understanding that she would return with her husband the following night.

When they returned the next evening, we immediately sensed a change. Her husband was full of anger, bitterness and hurt. He was struggling with the fact that Alexei was alive but not with them. He told us how the doctors had insisted that Alexei would die or that if he lived, even for a short time, that he would be severely brain damaged. The first they heard that Alexei was still alive was when they read an interview I did, accompanied with a photograph of Alexei in a local paper, stating that he was coming to Ireland for

that moment on it meant that Alexei had the security he needed and that Len and Chris could finally call themselves a family. We have maintained good contact with Irina, who has eventually managed to have a healthy little girl, Dasha. I have

life-saving treatment. After a while things calmed down and we promised to send them regular photos and reports of Alexei. Irina gave us gifts for him, a beautiful gold cross and chain and a toy, and asked Len to give them to him when it was appropriate to tell him about his Belarusian parents.

By the time we left Belarus we were elated but emotionally drained. But from

gone to see them many times, and looking at Dasha is like looking at Alexei. She even has the same stubborn, resolute streak. Irina drinks in every word and story about Alexei and still dreams of seeing him someday soon. Alexei has photographs of all his Belarusian family and often asks about them. Sometime soon he will ask to visit them, and Len and Chris are ready and waiting to take him.

Alexei is now 10 years old and has a marvellous life. He is happy, loving, confident, a brilliant footballer and always full of joy. He is bright in school and attends an all-Irish-speaking primary school and tells me that he wants to be an engineer, train driver or pilot. The only shadow in his life is the huge hole on the left side of his face. The only times I have ever seen Alexei really sad is when he thinks about his missing eye. Because it's such an obvious thing, Alexei has become the object of some very rude and inquisitive stares and sometimes even very hurtful comments. Despite the confidence he has, he is still vulnerable and says he would love to just be the same as everyone else. His dream has been about getting an eye. Michael Earley has been working with him over the past number of years and many discussions have taken place where Alexei has learned to understand how compli-

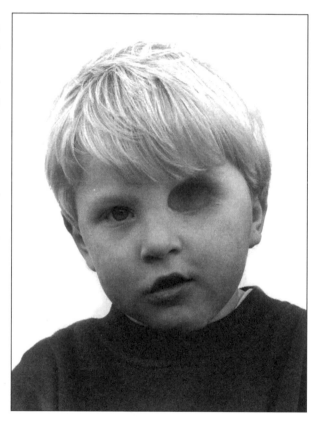

cated the surgery would be. To satisfy ourselves that we were doing everything to make life easier for Alexei, Len and I brought him to the US to see top surgeons and prosthetics experts. We didn't learn anything new, but it gave us the impetus to pursue what could be done in Europe, so

back we came to Michael, who decided that he was old enough to go through major surgery, which would implant further bone to try to improve the indentation on the left side of his face and to build around the orbit so that Alexei could be fitted with a prosthetic eye.

Even with the fitting of a prosthesis, Alexei would have a lot to contend with physically as he has only half a rib cage, which leaves his lungs vulnerable, as well as having some other medical issues. When Len brought him to see one of

Ireland's top geneticists, Mr Andrew Green, he told her clearly that Alexei is a wonder child medically, as by rights he should be severely brain damaged with a poor quality of life. Alexei's condition is rare, with only another five children in the world with the same diagnosis.

Michael Earley and his team agreed in 2004 that Alexei's skull had grown sufficiently for them to re-enter and to perform the first key step in rebuilding his face. The date was set for early September of that year, and while this was the decision that Len, Chris and Alexei had been waiting for all these years, as the time approached there was definite anxiety building up. As 28 September drew nearer, Alexei was being prepared by Len and Chris and he started to ask very specific, direct questions about the nature of the operation and what it would entail. The day before he was admitted to Temple Street Hospital he asked Len, 'Mam, is there a chance that I might die?' Alexei's question brought up a lot of unease in Len as she was, until that point, reluctant to face up to the dangers associated with highly invasive surgery. This was further compounded when Len was taken aside the night before surgery by the Registrar, who told her very clearly that there were serious dangers attached to the surgery. He explained that as they would be working very close to the optic nerve of Alexei's existing eye, there was a chance of damaging it. He went on to tell her that

they would also be working very close to the brain, which carried obvious risks, as well as the potential dangers of huge blood loss from the skull. Len phoned me after this meeting, reeling with the impact of what she had been told. I honestly believe that until that moment she had blocked out the reality of the kind of surgery Alexei would be undergoing and that she had succeeded, until that point, in remaining blissfully unaware.

Despite the knowledge that Michael Earley was, in the words of the Registrar, 'the best man for the job', the tension all that night was tangible. Alexei himself realised this was serious as he went through a battery of tests in the hours beforehand. Len stayed in the bed with him that night and held his hand right up to the moment of his being taken away by Michael and his team into the operating theatre. For the following six hours Len walked the streets of Dublin in a state of suspension until finally the phone rang and Michael advised her that Alexei had pulled through despite huge blood loss and the severity of the intervention. As Len rushed back to intensive care, she rang me crying on the phone with the good news. The following 24 hours were crucial for Alexei, but days later he was out of danger and back in one of the children's wards to continue his recovery. Len kept me up to speed on his development but nothing she said prepared me for when I stood at the bottom of his bed four days after the operation. Norrie and I had driven straight up from Cork and arrived at the ward with great anticipation as to how he would look. But we both were floored when we first saw him half lying, half sitting in the bed, his thin little body overwhelmed by his enlarged, swollen, bruised head, totally out of proportion with his body. Poor Norrie got such a shock that she had to leave the room. With great difficulty I stayed. I had a lump in my throat as I looked down at our little boy, but I knew crying wouldn't do anyone any good and I knew Alexei well enough to know that he was watching for my reaction. I told him how brave he was and how happy I was to see him. I held his hand and stroked his broken head as he haltingly answered all my questions.

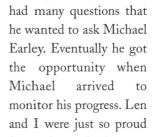

Len hadn't prepared us for how he looked because she had become so used to it, but Norrie and I were devastated. However, over the next couple of days we saw great improvement as the swelling went down and one by one a variety of drips and needles were taken out. He was in terrible discomfort, but rarely complained. His worst day was when he came out of intensive care and asked Len for a mirror, as he wanted to see what the surgeons had done. When Len gave him the mirror to hold, Alexei was inconsolable. He could hardly see with his one eye, as it was very swollen and blurry, but he saw enough to cause him to be upset and he said, 'Mam, I really want to cry but the tears won't come.' We later learned that his tear ducts were temporarily dried up from the operation.

Alexei is such an inquisitive boy and he had many questions that he wanted to ask Michael Earley. Eventually he got the opportunity when Michael arrived to monitor his progress. Len and I were just so proud of Alexei that day as he spoke 'man to man' with Michael, asking everything from 'how and where did you open my skull?' to 'when can I get my prosthesis?' Michael gave Alexei as much time as he needed and left him with the happy knowledge that he would have the prosthesis within three months of the operation. As Alexei prepared to leave the hospital, he made and wrote a special card for Michael which said, 'Mr Earley, you will always be very important in my life.'

Alexei's ability to battle against the odds has astonished many, but not those of us who know him well. He has what I call the 'X factor', which I believe he got from his birth mum and dad in Belarus, and which has to do with resilience and inner strength. Two weeks after the operation I had the opportunity to visit his mother, Irina, and during my time with her I shared the story of Alexei surviving

131

his life-threatening surgery. She hung onto my every word as we held each other's hands and between us shed tears of laughter and sadness – laughter at the relief of his survival and sadness because of how harsh life has been on children like Alexei. Irina took his photographs and kissed each one and told me how she thinks and prays for him each and every day. She is a stoic, very spiritual woman and is philosophical about what she and Igor had to do in order to keep their son alive. She is waiting for the day that he will return to visit them and get to know his Belarusian roots.

I came back to Ireland with gifts from his sister, Dasha, now four years old, and with the firm belief that Alexei has been very blessed to have two wonderful sets of parents all wishing the same thing for him – a healthy and happy love-filled life.

He and I have a special bond not only because I'm his aunt and godmother but because he knows, without ever saying anything, that there's a deep connection between us since the moment he locked eyes with me and challenged me not to turn away, to give him a chance at life. We have spent every Christmas together since then, and every Christmas when Len, Chris and Alexei arrive from Dublin, just before he nods off to wait for Santa, he asks me to tell him the story again of how I went on Christmas Day and brought him to Ireland forever.

He often asks, 'Why did Adi pick me?', and I say, 'Well, why not? You asked me when you looked at me that first time I saw you.' I would have been haunted for the rest of my life if I hadn't responded. I suppose, Alexei, it was meant to be; you were meant to be a big part of our lives and we are blessed because of it. Thank you.

In the year running up to the fifteenth anniversary of the Chernobyl disaster, I started to focus on how we could rekindle world attention on a tragedy that had disappeared from the international agenda and seemed relegated to the realms of history. I had been working with an English artist, Christine Simpson, who had seen and been very impressed by the work of a US-based photographer friend of mine, Paul Fusco of Magnum Photo Agency in New York. I suggested gathering other artists together to make a joint collaboration on the theme of Chernobyl. While we all agreed that it would be very productive and the work would be of a very high standard, we wondered what we could do with the resultant work.

The day we were having this discussion in our office in Cork, I mentioned how I had a very dear friend in the United Nations Department of Humanitarian Affairs, David Chikvazaide, and perhaps he could help us to get our fledgling exhibition accepted by the United Nations. You might wonder how we could possibly imagine that the world's most important agency would consider accepting an exhibition from such a small country, about an event that was fifteen years old – an exhibition which didn't even exist other than in our minds. Well, 'nothing ventured nothing gained' is one of my mottos, and with my usual enthusiasm I picked up the phone and rang the United Nations. David was at his desk and I blurted out how we had this concept for a Chernobyl art exhibition. 'David, would there be any chance that the United Nations would consider hosting this exhibition to mark the fifteenth anniversary?' David paused before he responded, and I couldn't believe my ears when he said, 'Actually, your timing is perfect on this, Adi. Word has just come from the office of the Secretary General, Mr Annan, that he would like us to put a special effort into marking this next anniversary of the Chernobyl disaster. So, what I'm saying now is, we're interested. Send me the details and we'll get back to you.'

With this good news ringing in our ears, we and the artists developed a plan for the next few months. We had no budget and absolutely no funds available for this project, yet we went ahead. It was carried by the energy and talent of all the artists, led by Christine Simpson, who was to become the curator of the exhibition, with the staff and I taking on the responsibility of dealing with organising the actual event, including the media, negotiating with the United Nations and all the various agencies. Paul Fusco was very excited about the project and threw his weight behind it from the beginning. It was a very ambitious project and it was a nightmare to organise. If we'd known then what we were to subsequently learn, I think we would probably have run a mile in the opposite direction!

UNITED NATION
FIFTEENTH
ANNIVERSARY
COMMEMORATI

'I appeal to Member States,
gov;ernental organisations and p
individuals to join with me in a p
never to forget Chernobyl. Togethe
must extend a helping hand to our f
human beings and show that we ar
indifferent to their plight.' *Kofi A
United Nations Secretary General*

The artists did extraordinary work. Christine isn't only a good artist, but also showed incredible talent regarding the actual development of the concept and making it a design reality. It involved some trips to the United Nations in New York, a zillion phone calls, e-mails and some very difficult meetings. About six months before the exhibition was due to open in New York, we had to face the issue of costs, which were escalating by the day. None of the money needed for such a major exhibition could be covered by the charity, as all our funds are required to be spent directly on our aid programmes, so we reached a stalemate situation, with bills mounting and nerves fraying as to where this money was going to come from. Then I had a brainwave. I was a great fan of our Minister of State for Foreign Affairs, Liz O'Donnell. She is one of the most forthright and intelligent politicians in Ireland, and at that point in her political career had responsibility for overseas development projects. I rang her office and was immediately told she would agree to meet me. I sent a briefing paper in advance about the exhibition so that Liz was well informed on what we were seeking. Within 20 minutes the Minister gave us her response – we could have the necessary grant towards making the exhibition become a reality!

Having crossed the Atlantic several times during the organisation of the exhibition, I finally moved to New York

with one of our volunteers, Mags Murphy, who had agreed to provide practical help with all the nitty-gritty problems we would encounter in the weeks leading up to the opening. Nothing prepared us for all the problems that beset us when we got to the States. Everything there worked so differently. We were so used to people in Ireland giving help whenever we asked for it; nothing had ever been refused to us. But in New York it was a very different kettle of fish altogether. While people felt it was 'admirable', 'very important' and so on, it didn't translate into the practical support we needed. So we turned to the constituency that we felt would be the most sympathetic – the Irish-American community. But while they were welcoming, it didn't translate into much practical help. When we probed as to 'why', we always got the same answer: 'Your work is really great, but the recipients aren't Irish, therefore it doesn't fall into our brief.' So even though we had this amazing exhibition about to display at the United Nations, part funded by the Irish government, a uniquely Irish project with an international dimension, it wasn't sufficient to encourage the Irish-American community to respond.

Mags and I didn't give up easily. We talked to everyone and anyone who would listen. At one stage we held a significant meeting with the international Irish aid organisation, Concern, and were told in very plain terms that we wouldn't make it in the US, based on our lack of a US-based administration. It took Concern 10 years to build up to the level they were at at that point and it had taken a lot of money, time and effort. Our friends in Concern even suggested pulling the exhibition from the United Nations and thinking of another venue. At that late stage, that would have been impossible. We were grateful for the blunt advice, but came away very despondent.

We had to face up to the reality that with only mere weeks before the opening, we would have to move heaven and earth to make it happen, regardless of the costs. I don't mean financial costs, I mean the mental and emotional costs, and they were very high, believe me.

Mags and I began to develop a plan. The exhibition arrived in the US. We hired the personnel to put Christine's concept into reality on the floor of the exhibition space in the visitors' lobby of the United Nations. We hired the company that would make us a stage for the event. We got in touch with one of the directors and composers of *Riverdance*, Bill Whelan, to ask if the *Riverdance* troupe would come down and take part on the opening night and he readily agreed. Kofi Annan was going to open the exhibition along with the ambassadors of Belarus, Ukraine and Russia. We were also very proud to have Minister Liz O'Donnell represent the Irish government.

Finally the day arrived, 26 April 2001.

The day started for us at six a.m. as we had to work on Irish time as well as American time. Lots of interviews and final arrangements had to be made before the close of business on one side of the Atlantic, and then we'd get ready for a full American day's work. We arrived at the United Nations to participate in a formal commemoration ceremony at the Peace Garden within the United Nations. It was a great way to calm and settle my nerves and to focus on what this was all about. I had almost forgotten the whole purpose of the day − to remember and remind the world about the plight of the victims of Chernobyl. The Under Secretary for Humanitarian Affairs, Mr Kenzo Oshima, himself a survivor from Hiroshima in Japan, gave a special and personal reflective speech as he tolled the Peace Bell.

There were further meetings after the ceremony, and in the afternoon I got my opportunity to speak at a UN conference. I was very aware of how important the audience was, and I made what I hope was an impassioned speech. With the speech done and dusted, we then went straight into the organised chaos out in the exhibition area. There were camera crews, reporters and photographers milling all over the exhibition, which was still being touched up with paint! I did an interview with Carole Coleman for RTÉ. As always, we prioritised Irish media, since after all, Ireland is our bread and butter. The caterers arrived and laid out the tables, the

Irish and Belarusian embassies kindly provided us with Irish whiskey and Russian vodka, the stage was finally ready and the public address system was ready to be tested, so all we needed now was the audience. Slowly they came until finally we had over 1,000 guests. Then the *Riverdance* troupe took to the stage. It was electrifying. I had asked them to perform the Russian-Irish section of the show and they were magnificent in their performance. The crowd was spellbound. UN personnel kept saying that they had never had an opening like it! Then it was straight to the speeches and formal opening. I was particularly proud of the speech made by Minister Liz O'Donnell on behalf of the Irish government. Liz is a fine speaker and went that extra mile in her compliments to the Project for all our work. The speeches were timed to the last second, so the opening event clipped along nicely and finished, appropriately, with Alexei, Raisa and Anna making traditional presentations of bread and salt, a common Russian custom, to all dignitaries.

So finally, after all the blood, sweat and tears (there were plenty!), the exhibition was opened and people began to walk around and see it for the first time. We awaited their reactions. Within 20 minutes the first people who had been right through it came up to us. They were speechless. They just embraced each of us, thanked us and went straight for a drink! They were deeply moved by the images.

People in New York had never seen anything like it. The realism of the exhibition seemed to take them unawares and unprepared. It was powerful to watch. In the following weeks, Mags and I spent every day escorting the general public and school tour groups around the exhibition. We marvelled at their reactions as people openly wept and hugged each other and invariably asked the most important question of all: 'How can we help?'

We have since developed an even stronger relationship with the United Nations, both in New York and Geneva, in our joint attempt to raise worldwide awareness on Chernobyl. We have been involved in a number of initiatives with the United Nations, including the launch in July 2003 of a new UN initiative called the International Chernobyl Research and Information Network (ICRIN). I had the privilege of making the keynote address at its inaugural meeting in Geneva and subsequently our organisation was invited to represent all the world's NGOs working in the Chernobyl area. This position has given a voice not only to those of us engaged in humanitarian projects, but it also gives a crucial place at the decision makers' table to those that are most affected by this tragedy. They are no longer left out in the wilderness and we hope that we give them fair and just representation.

In 2003 our organisation successfully applied for and received our official non-governmental organisation (NGO) status from the United Nations. In receiving this special status, our work has been recog- nised at an international level and it allows us to adhere more closely to the guiding principles of the United Nations.

Dr William Novik, Maryann deLeo, UN Under Secretary General Jan Egeland, Adi Roche

DUNCAN'S STORY

In the spring of 2003 I contacted the popular Irish broadcaster, Duncan Stewart, an architect by profession but whose talents and interests reach far beyond the boundaries of architecture. He is best known for his environmental television programming in which he examines the serious ecological issues facing the world today. In particular, he has examined the issue of energy, resources and the world's dependency on nuclear power. I was drawn to working with Duncan because of his track record on the nuclear issue and approached him to travel with a team of aid workers in April 2003. Duncan agreed to travel with us and brought along a camerawoman, Roisin Leggett. This trip was pivotal to Duncan's decision to join forces with us to make a documentary on the consequences of the accident while also looking at what the solutions may be.

On our return we looked at what footage we had and realised we had the makings of a very strong film. We planned to return to complete further filming in the autumn of 2003. At that stage we had persuaded Duncan not only to help us make a documentary, but to act as our architect on an Irish government-sponsored building project. We were joined on this trip by the Irish photographer Julian Behal, who had offered his services for free. On our first day we ran into a problem, as our Belarusian cameraman had a very serious car crash just before we arrived, which meant that Duncan had to take over the role of cameraman. This setback didn't dampen our enthusiasm, though, and off we went to start filming. On day three we had settled into the semi-deserted small town of Bragin, a strange place situated on the edge of the 'purple zone'. It's difficult to express in words all that happened during that time in the radiation zone. We existed in some kind of twilit lost world which is part deserted, part inhabited by all sorts of 'odds and sods' from as far away as the Ural mountains and the various 'Stans' (Uzhbekistan, Kazakhstan, Tejekestan). People who were fleeing wars in their own country have chosen to live in the deserted houses of the contaminated zones.

As this entire area is heavily contaminated, our plan was to stay overnight, enter the inner 'exclusion zone' and film right down at the reactor site itself. We set off early, taking driving shots where possible of rural life in the zones.

On the outskirts of Bragin we found ourselves in a place called Dublin! It turns out that in medieval times, pilgrims from Ireland passed this way to and from the Holy Land. It appears that not all the pilgrims went home, for some settled in what became the new Dublin and integrated with the local people. We were intrigued by this story, which was further confirmed when we found the only thatched-roof houses in the whole of Belarus. As this is a traditional form of roofing in Ireland, it added to the possible authenticity of the story.

We continued to the Belarusian border point with the Ukraine and, as there were no border guards present, we proceeded into the 'no man's land' area between both countries until we arrived at the Ukrainian border post, which was heavily manned. While Natasha, our translator, was taking care of paperwork, I went wandering around the post and found out that this was the place where people are scientifically tested for radiation when going to and fro across the border. I told Julian and Duncan that we should film this, as it was a unique opportunity to show the daily routine of living in a radiation zone. The scientists were only too willing to allow us to film workers going through, but wouldn't allow us to film the vehicles being checked by hand-held Geiger counters. Having completed the necessary paperwork, we proceeded into the 10 km 'inner exclusion zone'. Almost immediately, looming ahead of us, we could see the Chernobyl nuclear power plant dominating the skyline. We were nearly there. Our first stop was at the town of Chernobyl, which no longer houses the families of the workers and is now part of the scientific laboratory. We stopped to film and pay our respects at the monument to the fire-fighters who died, then went directly to the reactor site itself. My impression of the site is always the same. There is something about this grey,

enormous, apparently innocuous building that sends a shiver down my spine. It looks so innocent, yet holds so much destructive power.

We were advised by our guide that since 9/11 film crews are no longer allowed to move freely around the site, as it has become a 'grade A' target for possible terrorist attack – a chilling thought, considering its lethal potential. We filmed as much as we were allowed to at the site, then went on to the custom-built town of Pripyat that had housed thousands of the plant workers and their families. It was a sorry, sad place. You could see where once it had been a thriving community with all the amenities of modern life – a cinema, swimming pool, playground. Now the playground's rusted Ferris wheel makes a grinding, eerie noise that breaks an uneasy silence. There was something about it that had the effect of both attracting and repelling you at the same time. I had the same sensations as we came down to the banks of the river Dnieper and saw abandoned barges; it was like witnessing a modern-day Pompeii.

As the light started to fade, we quickly decided we'd better finish and get back to do one last shot that Duncan needed on the Belarusian side. We sped towards the Ukrainian border with Belarus and proceeded to the spot that Duncan had chosen earlier for what he called his 'mood shot'. Our driver brought us to what was once a beautiful village, now deserted, its

houses showing signs of neglect and the gardens overgrown, now like the heart of darkness. These quaint timber houses had absorbed huge doses of radiation, and for those who had lived in them it had been like living inside a mini reactor because of the porous quality of timber. Julian and I were set the task of keeping tabs on the Geiger counter levels and to get all the gear together to assist Duncan's filming. Just as we were standing in what was once someone's apple orchard, we switched on the counter. We both stared at the reading. The counter had jumped enormously since our previous reading just minutes earlier. Holding our breath, we ran from that deceptively lovely garden and found a safer place in the garden next door. We were afraid to go indoors, although psychologically we thought we would be safer. For the first time we acknowledged each other's fear. What we had just experienced was a phenomenon known as 'spotty contamination'. These 'hot spots' are everywhere, invisible except to the sensitivity of a Geiger counter. It's apparently impossible to decontaminate such patches of heavy contamination.

While Julian and I were recovering for a brief moment back at the van, Duncan could see the entire Chernobyl site in the background and wanted to get the shot. What happened next will be printed on my mind forever. As time wasn't on our side, what with the fading evening light, Duncan jumped out of the van and ran ahead to get the best vantage point for the shot. Julian, Natasha and I were to bring the equipment and cameras. I called out to Duncan. When he failed to respond, we went to a spot where we thought he had gone and it was there, at the base of a rotten tree, that we found him lying unconscious. At first we thought it was a joke, but when Julian took his pulse, we knew something was wrong. Julian described his pulse as 'thready'. This expression threw a shiver up my spine, as the only time I had heard that expression before was when my father was dying. We assessed the situation and gathered that Duncan had climbed the tallest tree, not realising that the inside was rotten from radiation. The tree imploded as soon as Duncan stopped climbing and then, on his fall to the ground, he banged his body severely against an old building beside the tree and landed heavily at the base of it.

So there we were, in the middle of this odd world trying to make a documentary, with Duncan in a rapidly deteriorating condition. Without access to a phone or ambulance, we were forced to carry him to our van. We drove him as fast as we dared to the Bragin village hospital. Duncan was the first foreigner to be treated there. The doctors were terrified when they saw his deteriorated condition, and of course they were extremely worried about the consequences if he should die, in case they would be held responsible. There we were with Duncan in a near-death state, unable to contact the outside world. There was no outside phone in the hotel, on the street or in the hospital. On the first night, when they told us that he would die, we had to cope alone and make decisions. The responsibility and fear were huge, the feeling of aloneness almost overpowering, and the sense of pure and utter panic knowing that his wife and family were sleeping peacefully in Ireland, oblivious to the reality that Duncan was dying, was unbearable. I saw Duncan hover between life and death. He aged about 20 years. His colour was the grey/yellow of death, his body was cold, his breathing slow and laboured and he was slipping in and out of a coma. His condition was so deteriorated that without knowing if he was a man of religion or not, I had to decide to give him the last rites. I felt that if he were to die and was a Christian it would comfort and help him 'on the way', and if he wasn't, then it wouldn't do any harm anyway. I whispered an Act of Contrition into his ear as part of the last rites. Just in case.

Duncan didn't die that night, but flitted between life and death. Throughout the most critical of times he showed great courage. Being with him as he lay there brought back memories of being with my father in his last hours of life. Duncan's appearance and condition seemed the same. When he was conscious, I had to explain to him that he had to have emergency surgery and needed his agreement. He asked me, 'Am I dying?',

and I answered, 'Yes, I think so', and he gave his consent. Never once was he self-pitying and always showed incredible dignity. Even when we had to keep slapping his face to keep him conscious, he never complained. When he was capable of uttering some words he said, 'What about the documentary?', 'What about the lads on the building job?' or 'I've let you all down.' He never had a thought for himself. I started to take notes on things that he uttered, sometimes in his sleep and sometimes when he was lucid. It was as if he was giving us a final message.

The hospital staff in Bragin saved Duncan's life. The ventilator machine that he needed to stay alive was a gift from the Swiss government. It was thanks to humanitarian aid that Duncan was kept alive for those vital days after surgery. Despite difficult telecommunications, I was finally able to make phone calls back to Ireland, and the day after the surgery managed to get a call through to Duncan's brother, Ercus. I felt that Ercus, who is a barrister, would be level-headed enough to help me to make decisions on what to do next. He became my lifeline as he remained focused and calm. Only when talking to him could I allow myself to release the tension and worry I felt, as I had to keep a brave face when with Duncan. Ercus got one of Ireland's top cardiologists, Mr Con Timon of Saint Vincent's Hospital, to give me a list of questions to ask the local surgeons and doctors, and based on the information I gathered he was able to advise both Ercus and I about what to do next. We were all agreed that Duncan should be brought back to Ireland as soon as possible, but at that stage he was still in an unstable condition. There were so many things that could go wrong for him, and if we had made the wrong decision at that early stage it could have been fatal. We knew that to move him to a better-equipped hospital would be safer, but we were also acutely conscious that he could die in the process of being moved. The nearest hospital was at least two hours away by ambulance, so we had to wait until he became more stable and had a better chance of being stronger for the journey. While I went back and forth between Ercus, the cardiologist and the local doctors, our office was frantically trying to find a way of getting Ercus and Duncan's son, Marcus, out to Belarus. I wanted them there in the event that things deteriorated. The burden was too great to be carried by Natasha, Julian and me.

While this was all going on, another unforeseen problem hit us fast and hard. The state security bureau, still known as the KGB, arrived at our hotel and informed Natasha that we were all going to be arrested and that we had to appear in court the following day. In the list of people to be arrested they included Duncan, despite their knowledge that he was seriously ill. We were told that he

would still have to appear in court! Eventually he got a dispensation and Natasha and our driver, Slava, went to court for the first time. When we questioned the reason for this arrest we were told that we had crossed the border into Ukraine 'illegally'. This was untrue, as all of us had been issued with visas in advance of our trip and all had double entry visas in and out of the Ukraine. What had happened on the day we entered the exclusion zone was that there was no one manning the Belarusian border point. As people rarely cross the border through the exclusion zone, all we could suppose was that the guard had rambled off somewhere, thinking no one would be crossing, and left the border point unattended. Despite our paperwork being in order and despite our explanations, we were told in the court by the judge that we would have to appear at a further stage for sentencing and to be fined. There was also mention of a prison sentence.

So in the middle of conversations with Ercus about Duncan, I had to ask him for legal advice. I was getting concerned about the way things were going and we were now being watched by the KGB all the time. He told me to try and stall for time by saying that our legal representative would be arriving within two days and would represent us at the next hearing. When I informed our KGB man of this, he seemed to take extra note of this development, particularly when I told him that Ercus was also an international lawyer and arbitrator of repute. We managed to get our court date put back until his arrival, but we knew we were by no means finished with the courts. Ercus contacted the Irish ambassador to Russia and Belarus, Mr Justin Harmon, who in turn contacted the Department of Foreign Affairs in Minsk, and this intervention strengthened our case. Despite having friends in various government departments trying to get this absurd case dropped, we were constantly aware of the KGB man following our every move.

Two days after Duncan's surgery, Natasha and I were eating cold creamed rice out of a tin in the corridor outside Duncan's room. Suddenly there was a flurry of activity, and marching down the corridor came the surgeons from Gomel Regional Hospital. They started to prepare Duncan to be moved to their hospital. There was some heated debate between the surgeons and the local doctors, who felt he was too unstable to be moved. The 'big guns' won the day and Duncan was moved onto a trolley and taken out to the waiting ambulance. It was all happening so quickly that there was little time for discussion and my opinion wasn't asked for one way or the other. In lieu of having family, I felt I was Duncan's representative and asked to sit with him in the ambulance. I was initially dismissed with great disdain, but I held my ground

and was eventually permitted to travel with him. Duncan was delirious, confused and distressed for a lot of the journey and I just prayed to God that he would make it to the hospital. When we arrived, I expected there to be a team to help us, but there was no one there. Half an hour later, Duncan was taken to intensive care. On arrival there I stood at the door to allow the medics to move Duncan from the trolley to the bed, as the space in the room was very narrow. Disaster almost struck when one of the young medics pulled the wrong lever, which had the effect of collapsing the stretcher. In milliseconds Duncan was slipping, head first, onto the tiled floor, with glass bottles of various medicines attached to him crashing and splintering on the ground. Fortunately he was caught in time and laid on the bed. This episode worried me and I saw the effect it had on Duncan.

The only relief I felt was in the knowledge that Ercus and Marcus were arriving within the next five hours. Standing at the side of the road at midnight that night and seeing their faces appear in the back of a Lada car was just the best sight! They immediately wanted to see Duncan, and after much negotiation we succeeded in getting them by his bedside. I felt like a bit of an interloper during their reunion, but it was a privilege to be there. All three were quite macho and almost bantered with each other as they searched within themselves to cope with the drastically altered Duncan. The room was filled with unspoken concerns and emotions. I could see that both Ercus and Marcus were deeply moved and worried. Duncan told me later how he remembered that moment. He said, 'I thought I was dead. I felt that Marcus and Ercus were looking at me laid out in a coffin.'

The next day we met with a variety of consultants to discuss what was to be done. In between discussions with doctors and insurance companies, Duncan began to improve. I knew there was a change when he announced he was starving! I told him that when the food trolley came around that he should ask for some potatoes and vegetables and not to have anything too heavy on his system. With that I left him content with the thought of food. Imagine the shock when on my return the next morning I discovered he hadn't received any! When I enquired about this from Dr Alex, who spoke good English, I realised that it was a delicate matter. I had to read between the lines, and after a few minutes it dawned on me that there was no food for patients in the hospital and that families had to bring their own food to loved ones. With that I sent Marcus out to the ambulance and got him to bring me a tin of tuna fish, some Irish butter, cheese and brown bread and some juice. Dr Alex had kindly received permission for us to use the staff room to prepare a meal for Duncan, which I

eventually brought to him. He was unable to feed himself, so using a teaspoon I fed him the tuna and buttered brown bread. He loved it and said it was the best meal he'd ever had and quickly demolished the entire plate.

That night I felt I could finally speak to Agneta, Duncan's wife, and got her on the phone. I told her everything, and having done so, I then started imaging all the things she must be going through – the disbelief, the grief, the fear, the not knowing, the helplessness. I had thought so much about her and how she must be feeling; it was a relief to speak to her and let her know that her husband would survive after all. That night I slept better than I had for days. It was like a load had slipped off my shoulders and the burden was now shared.

Duncan was still not out of trouble and slipped in and out of consciousness most of the time. When he had lucid periods they lasted only for short minutes and we often had a laugh to keep his and our spirits up. Duncan used to say things like, 'Look, if I die here, right, don't bother trying to get me home, it's too far, just throw me out in the forest. I'm a nature person and I'll be grand!' Another time I was in dispute with the authorities about not stripping him on the runway in minus five degree temperatures, and when they pulled the blanket from his body he started singing a very rude Dublin song! There were extraordinary moments in the

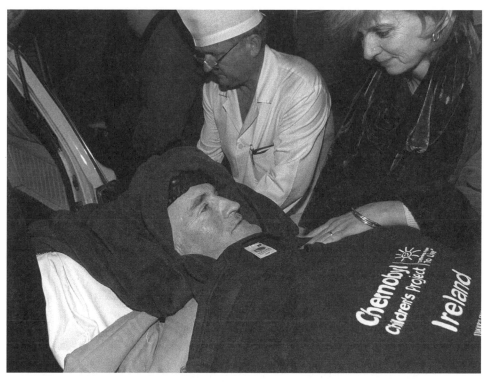

midst of the insanity. Duncan was also very proud of getting almost three litres of Belarusian blood; he'd say, 'Well, I'm half Belarusian now, none of ye can say that!'

Duncan has survived against the odds. He has stunned the medical profession, not only because he is alive, but by the speed of his recovery. I'm convinced that he must have mountain goat genes and a very strong will to keep on living! Throughout that time we tried to keep his spirits up and to take his mind off dying. It was an exhausting, almost over-whelming task, as our spirits and hopes were also low. Somehow we pulled each other up.

Two years later, Duncan is still recovering from his accident but hasn't wavered in his commitment to the victims of Chernobyl. When unable to travel to the region himself, he has nominated his son, daughter and nephews to go in his place and carry on the work. He has also rallied the support of his wife, Agneta, to the cause and she now organises an annual Chernobyl Heart Ball in aid of our work. Agneta, I believe, would have had every right to close the door in my face and never want to hear the word 'Chernobyl' again, but instead she has chosen to join us with great enthusiasm and support and has become a dear friend.

In November 2005 Duncan returned to the Chernobyl region with a film crew from RTÉ with a view to retracing his steps of what happened before, during and after the accident. The documentary we had been making now changed to become the story of Duncan's life-changing experience resulting from his accident. The film shoot was to prove to be a cathartic and pivotal turning point for Duncan. For the intervening two years he'd had little or no recollection of what had happened, and little by little memories started to dart back to him prompted by revisiting the locations.

The documentary to mark the twentieth anniversary of the Chernobyl disaster is a powerful reflection of how Duncan's indomitable spirit came shining through against the odds. His reluctance to make a programme about himself was something that almost prevented him from doing the RTÉ film and he only agreed when he felt confident that he could dovetail the story of Chernobyl on the back of his personal tragedy. He was deeply moved by how the money raised because of his accident had saved the lives of so many children suffering from the Chernobyl Heart condition, and when he met a group of the children and their parents, who came to thank him, he said, 'I would have died happily with the knowledge that through my death so much life would be saved.'

Duncan meeting the group of children who wanted to thank him

148

DOCUMENTARY FILMING AND THE MAKING OF AN OSCAR

After my first visit to Chernobyl in 1992, I became haunted by the sounds and images of what I had seen. I fretted about the future of the people and their children. No matter how much my black and white photographs could portray of what I'd seen, it just wasn't enough to express the enormity of the problem. I was left with the question of how to evoke a world response. What would it take? With these questions in my head, I wrote my first proposal for a documentary and the seeds of what eventually became a major film began to gestate in my mind. Luckily I met very encouraging people, and each contact led me to another until eventually I got a phone call from a woman called Isobel Stephenson who introduced herself as a film editor. She was fascinated by the story and within a few weeks I was discussing the making of a documentary with an independent film company. The fruit of this collaboration was to become the first documentary in the English language about Chernobyl.

The name of the company was Dreamchaser and they were so enthusiastic about the potential of the film that they were also willing to pay for it! Frantic work went on for a number of weeks and an approach was made to Ali Hewson to become the presenter of the film. Ali is deeply concerned about environmental issues and was ready to make a stand on the question of nuclear accidents and their consequences.

A wonderful, professional film crew was put together and we left Ireland to start three weeks of filming on 19 April 1993. At that time in the Soviet Union, secrecy was still very much in place, particularly regarding the aftermath of Chernobyl. Looking back on it now, I realise just how brave and courageous we all were. Our innocence carried us a long way, I think! The programme was shot on 16 mm film by cameraman Donal Gilligan, an absolute artist in his field. So, combined with Isobel's brilliant editing, Dan Birch's sound and Liam Cabot's producing, along with the directorship of Gerry Hoban, the most amazing film was made. It was a classic, made with a real feel for Russian history and in a somewhat Russian style of filmmaking. *Black Wind White Land: Living with Chernobyl* went on to be screened in several countries around the world. The international launch of the film was in Dublin and attended by none other than the entire band of U2 (Ali happens to be married to front man Bono). Having them put their weight behind our work has led to huge support amongst their fan base.

From the success of the first documentary I got a real taste for film-making. I have subsequently been involved in a number of films about Chernobyl, for BBC, RTÉ and many others. It was something I discovered I really enjoyed doing. I particularly love the research, production and developmental side of the work, where one can explore and give voice to an issue, both literally and figuratively.

A couple of months after returning from the United Nations exhibition in New York in 2001, I got a phone message that someone in the US called Maryann deLeo wanted to make contact about making a film. Maryann was a filmmaker and had seen our exhibition at the UN. She was so moved by it that she wanted to try to make a film about Chernobyl. On receiving Maryann's CV, I knew she was genuine, and I sent her an e-mail welcoming her request. Despite her enthusiasm, I had to give the proposal serious consideration. The clinching factor for me was when Maryann said that the essence of the film would be through the eyes of the charity and would be a portrayal of our extensive work in the region. Obviously this was key to our co-operation, as the film would be a way of obtaining much-needed funds for our work, and Maryann said we would be able to use the finished film for fundraising. In subsequent e-mails I felt we got to know each well and that she had a good feel for the subject. I invited her to come and join us on a mission to Chernobyl that was happening a couple of weeks later. Her film proposal had the backing of US TV giant HBO (Home Box Office), so with a ready-made outlet

for the film, it made perfect sense to go ahead. I looked for a contract between us at that point and was refused, but being naïve about such things I allowed my enthusiasm for the filming project to take precedence over my judgment.

The preparation for the trip was intense, as we had to be careful about where and what we would request to film. Everything was done through the Project, from organising visas, accommodation, transport and translators to arranging the special passes required to enter the exclusion zones, as well as deciding and

putting together the entire filming schedule. Not only were we bringing Maryann to film, but we were also working at the same time with an Irish television crew from TV3, Brian Daly and Paul Deighan, which meant that two film crews had to be satisfied with footage, interviews, etc.

Over two trips we travelled the highways and byways of the radiation zone, seeking out stories and places. We slept in mental asylums, orphanages and strange hotels. We ate tins of tuna fish, crackers, chocolate and sweets, and a real treat was a cup of instant soup along the way. We all got closer as the trip went on.

Maryann and her friend Angie, along with Natasha and I, always slept together, sleeping bag by sleeping bag. There were many nights when we needed the comfort and warmth of each other after a difficult day's work.

On our first night we drove down a minor road, and when we finally reached our destination, Vesnovo Mental Asylum, I quickly realised why it was physically placed at the end of the road. The message to the children housed there was to become crystal clear – that they had reached the end of their road, the end of their lives. The message to the outside world was to say that this place didn't exist.

I remember our first night there and how we reacted to the bleakness of the building, the cold, dark corridors and the strange smells. We arrived when it was dark and all the children were sleeping behind locked doors. We went to the toilets in pairs, as they were quite a distance from our room and down a dark corridor. We eventually fell into an exhausted sleep. We were woken up to the sounds of our locked door being banged and rattled, as children had heard that visitors had arrived. When we finally opened up the door, there was a convergence of children of all ages and sizes clamouring for attention. Some caught our hands and indicated that we should follow them. Judging from the all-pervasive smell of food, we knew we were being taken to the dining area. What a sight to behold,

150 children all shouting together for their food! The smell of what was about to be served was very strong, in fact almost overpowering, but this didn't appear to bother anyone except the visitors.

After breakfast Maryann took her camera and I brought her around to the first unit that was home to the smallest children. As soon as we opened the door, we were rushed by about two dozen children, all craving to be lifted up, held close, loved. Those that were unable to catch our attention were crying, so we had to quickly change children and give as many as possible a good hug and a quick swing in the air. We did our best to bring some calm into the room, a fairly impossible task as the kids were wild with excitement. It was only when I reached into my bag and found a bottle of bubbles and started blowing them into the air that an almost reverent hush fell over the children. They were mesmerised! Their faces were transfixed with wonder. Then the bubbles started to burst and disappear and there were tears all round. But once they saw how quickly new bubbles appeared, we were back to smiles again.

But not every child smiled. Sitting alone on a bench was a boy who looked about seven or eight years of age. He had very thin, weak hair and a very pale face. I looked at his hands and saw the reason for his sadness. His hands were a mass of swollen, purple-yellow scabs. I asked Natasha to go and talk with him. We

151

found out his name was Sasha and that he was actually 14 years old, even though he was the weight and height of a child half his age. He indicated that his feet pained him too, so we carefully took off his boots and socks to see what was the matter. Each movement caused him terrible pain. We were soon to see why. His feet were similar to his hands, but much worse. The soles of his feet were completely broken out, oozing a clear liquid, covered with large scabs. We were shocked by his condition, and on talking with a team of dentists that we had working with the children's teeth, who also had good medical knowledge, we concluded that he was at risk of losing both his hands and feet as they had the appearance of being gangrenous. Maryann had the difficult task of filming all of this, and my only way of dealing with Sasha's suffering was to try to find a solution. Our top dentists, Mr Jones and Dr Patrick Quinn, said that as soon as we got to a phone we should try and get Sasha accepted into their local general hospital in Tralee, County Kerry. Natasha and I set about getting his paperwork organised. This task was no mean feat because even just to get something like a passport we needed to find transport to travel 60 miles to the nearest town that had photographic equipment and then the long haul of getting the various official stamps would begin.

In the meantime, we continued filming all the other units. The story wasn't to get better, but worse. We went to the high-dependency unit and were shocked by the poor condition of many of the children as well as the condition of their beds, some of which were held together with timber pieces and twine. The children's faces were covered with masses of flies, but none had the strength to brush them away. Many of the children were emaciated, with paper-thin, grey-white skin pulled over seriously contracted bones. They just lay there, many of them soaked in their own excrement and urine, on very dubious mattresses. It was obvious on examination that many of the children hadn't been out of their beds for years. Living like this had rendered them severely contracted and totally immobile. It was appalling. I wanted to weep, to scream, to run away, but I couldn't. I was rooted to the spot.

Eventually I pulled myself together and started to ask questions, keeping in mind the sensitivities of being seen as foreigners and interlopers. I was careful in my questioning not to cause offence or to appear to be judgmental, even though I was. I quickly learned how much pressure the staff working there were under. They had no qualified nurse or doctor or enough childminders for the number of high-dependency children in their care; nor did they have access to sufficient clothing, bed linen, beds, nappies or medicine.

In between bouts of filming we fled back to what we now called our 'cocoon' or 'refuge', to comfort each other and to come to terms with what we were dealing with. For a change, we worked outside, filming the more able-bodied children working the land. As all institutions have to be self-sufficient in terms of food, the children had to work the fields to ensure a supply of food for the entire institution. During that visit and subsequent visits I was to witness the children knee deep in snow with no proper clothing, gloves or footwear as they bravely worked with the most basic farming tools. Their hands were often purple with the cold, their feet both purple and wet.

We stayed in Vesnovo for just one more night and moved on for the next part of the story. We travelled deeper into the zone each day, getting closer and closer to the exclusion zone, and started to prepare for the big trip down to the actual reactor site itself. The day finally came. We stayed in Bragin on the edge of the exclusion zone. Freezing with the cold we all slept badly, I suppose if the truth were told, we were also scared in anticipation of what lay ahead. There was a rising tension but nobody spoke about it, even though we were acutely aware of it. We kept our conversation technical, about what needed to be done, to ensure that the work was carried out with maximum speed and efficiency and with the least amount of risk. None of us had any desire to hang

around the 'purple zone'. In preparation we had agreed between both crews to take no risks and that none of us would jeopardise the health and safety of the others. Our Geiger counter would be permanently switched on. We were to wear full protective clothing. We agreed that under no circumstances would any of us remove any of the clothing or masks.

That morning we left at 6.30 a.m. to meet with one of the exclusion zone special police. He briefed us on what to expect and advised us to be ready at any time to stop and film. Speed, he said, was essential. We would be timed for a certain period and if we went over that time we would seriously endanger our health. Believe me, there was no argument with that logic! We travelled in our van and as we progressed we noticed subtle changes along the way, particularly once we passed the special border post. This post denotes the line of demarcation between what is considered reasonably safe and the area that is deemed to be extremely dangerous. We entered at the border point outside Bragin after the border guards introduced themselves as our hosts to the zone. They told us that we must have luck on our side, as it had rained the previous night, and therefore the threat from airborne radioactive particles was reduced. We felt suitably consoled.

We saw our first radiation warning signs and large public notices telling us to 'STOP! Do not pick flowers, berries, mushrooms, or do not hunt, do not touch the earth' – basically everything bar telling you to stop breathing.

The road so far was still in good condition despite years of neglect, but all along the way we began to see the first casualties. We saw the deserted villages and towns, overgrown farmland and overgrown, neglected gardens. As we rattled along observing in silence, I flicked on our Geiger counter. While it was showing fast increases in the levels of radiation, I wasn't unduly alarmed. Yet. The road deteriorated as we went deeper and we encountered many fallen trees, rotten from radiation exposure. This road runs the gauntlet between contaminated woods and newly ploughed radioactive land. We finally came upon the last border post, which would lead us into the inner exclusion zone of 10 km around the reactor. At this point we needed to change out of our clothes and don our protective clothing, consisting of dust suits, masks, overshoes and gloves. Underneath, we wore old clothing and footwear which would have to be deemed radioactive afterwards and carefully disposed of. We all felt very strange as we looked at each other. As well as being scared, we also saw the funny side of our situation, as we looked like the Tellytubbies, all puffed up in our white gear.

Having cleared our entrance into this final zone, we sped off in what was now a failing light. We concentrated our efforts on getting to the reactor site before we lost the light. Otherwise all that risk would have been for nothing. We pulled up by the bridge, then had to run up a steep pathway to get to the railway bridge, and from this vantage point we saw it. There it was, looming in the dropping light, this dark, seemingly innocuous series of buildings, against the purple-red sky. Over the Dnieper River lay the home of so much destruction.

My job was to monitor levels of radiation, and at some point as we ran up the steps onto the tracks, the Geiger counter went crazy, absolutely mad. It seemed unable to count fast enough. It no longer leapt forward in single digits, but in multiples of 100. Then the Geiger counter locked, unable to register the levels of radiation in the environment. I gestured from behind my mask to both crews to get going, fast, that it was more dangerous that we had anticipated. When I showed the guard the Geiger counter he got very anxious and kept telling us urgently in Russian to '*paslei, paslei* [hurry, hurry], this is not normal.' The pressure on both crews was enormous. Their dilemma was to prioritise the best shots while wanting to film everything, but with rising radiation levels it just wasn't going to be possible. Neither crew was able to film all they wanted, as we were now driven by the spiralling levels of radiation as well as our fear for our safety.

On returning to our van we quickly

sped out of the area and only felt relief after we dumped our now-radioactive clothing and donned our 'safer' personal clothes. There was silence in the vehicle as we all contemplated what we'd been through. I leaned my head against the fogged-up window that couldn't be opened because of radiation dust, numbed, sweating, terrified underneath my suit, which really gave little protection except against dust. I began to question my very sanity. What was I doing in the world's most radioactive environment? I felt engulfed by a strong sense of panic. I couldn't breathe. My heart raced and felt as if it might burst. The needle of the Geiger counter was still rising. It actually locked twice, incapable of registering the levels. Chernobyl's hot breath was everywhere. My fear was almost overwhelming. Feelings of sinking deep into a black hole rushed forth like a torrent. I screamed silently. My mind was grappling with what had brought me to this point in my life and carried me to the heart of Chernobyl along dusty radioactive roads. What pulled me back from utter panic were the crews in my care, the responsibility for their safety as well as their work. They had to be able to tell the truth about Chernobyl. Even though my earlier bravery had evaporated, I managed to recover enough not to show my inner turmoil. I looked around and saw each of the crews locked in their own private worlds. I felt as if we were in a state of suspension between two worlds. I frantically searched into the darkest recesses of my mind for consolation and thought of my mother, who had promised to light a candle for our safety every day we were away. Images of Mum, Seán, Len, Alexei and others came to console and sustain me. Prayers and words of laughter squeezed their way in between the blackness.

With the backdrop of lifeless fields, I thought of the despair of the peasant farmer unable to cultivate his land. I thought of the grief of the mothers watching their children fall ill and die. I thought of the silent sorrow of the old people forced to leave their native places they held sacred, places where they had spent their lives and where they left the graves of their loved ones.

The next day we had to return to the exclusion zone to film the deserted farmland and villages. Our special policeman returned and brought us through the same process. Clothing had to be replaced by protective gear, radiation borders had to be crossed again. We urged each other forward with an even deeper conviction about the story we had to record. We drove on to what was once the largest cattle farm in Belarus, the pride of the farming community. We filmed empty cattle stocks and recorded the sounds of rusting gates that once protected the cattle. Across the road lay the place where the workers had once had a blissful life. It was called 'Sunny Town', as it was in a beautiful place and was well designed, with what had once been a thriving, bustling community. We filmed at the kindergarten, the schools, the local theatre and cinema, the once-lovely apartments, finding remnants of life all along the way. In the school we found all the children's shoes, all neatly lined up under their hanging coats of all sizes, colours and shapes. We found their exercise books, read their essays about their lives, wandered through their classrooms and saw the lessons being studied written up on the blackboards, their final lessons, final studies in that place. We found children's toys, their dolls, their childish drawings, their footballs and playhouses. Everything was neglected, thrown about by the vandals who raid the deserted villages for booty to sell to unsuspecting customers in the markets of Poland and Russia. We found kitchens full of utensils where meals were being prepared on that fateful day in 1986 when their lives were changed forever.

Just as we were regrouping we heard the unusual sound of an engine racing. The policeman was immediately on alert. He took his gun out and held it at the ready. Within seconds we saw a very fast-moving car and trailer driving madly in front of us, carrying what looked like the contents of a house and all the floorboards. The policeman was angry because these 'bandits' enter the zones illegally and rob the towns and villages of anything of value to sell. Not only is this dangerous, as everything they rob is radioactive, but it's also like desecrating a graveyard. A bit shaken, we moved on to film some of the most beautiful but dilapidated ancient villages and farms.

We eventually agreed to move out of the zones. We'd had enough of death and destruction for one day. Leaving the 'no man's land' of the zone, I began to withdraw into myself. Within minutes of leaving there we came to a village almost identical to the ones we had just left. It was the same except for one thing – this village had life. Children played on the dust road; old people chatted outside their homes, drinking tea and smoking pipes; hens pecked around the grassy verges. Gaggles of geese held up the traffic as they crossed the road. I saw a woman hanging washing on the line and thought of the abandoned washing on the lines of the houses in the zone. I saw a little girl stroking the doll in her arms and flashed back to the doll on the floor in 'Sunny Town'. I couldn't stop my silent tears flowing at the pain of it, the suffering of it all. I cried for myself too.

Over the following days we visited endless hospitals, homes for abandoned babies, orphanages and families with sick children. We spoke with and interviewed scientists, doctors, mothers, the elderly – basically anyone we felt could contribute to the telling of the story.

In the midst of all this, we had moments of pure camaraderie where I would sing Maryann and Angie to sleep or we'd play a host of silly games to keep our spirits up. By the time we arrived back in Minsk, it felt as if we'd been away for several lifetimes and been through a dozen war zones. With our heads buzzing with a thousand images and feelings, we eventually got back to Ireland. The TV3 crew had the daunting task of not only producing one documentary but two, all within a period of seven weeks of broadcast. Brian not only had to do the editing and voice-over but also had to appear every day as the daytime newsreader. He missed the entire Christmas holidays as he had to work frantically around the clock with the editing. Finally, in January 2002 the two programmes were aired. They were very powerful. I felt that Brian and Paul had not only reflected the true scale and consequences of the accident, but also showed a very special empathy towards the people affected. Their programmes also paid a wonderful tribute to all the volunteers in the Chernobyl Children's Project International as they showed the vital work we do in the region.

Maryann returned to the US and over the following year or so worked on the documentary, which was to reach a whole new audience in America. I was so excited at this prospect, as Maryann wanted to show the American people not only the story of the tragedy, but also the story of Ireland's unique response. However, at a lunch meeting with Maryann during a visit to the US in September 2002, I was shocked to hear from her that the focus for the entire filming project was to change. HBO, realising the power of the footage we had obtained, decided they no longer needed the story of the charity. The impact of this, Maryann told me, would be that we would no longer be a central part of the film. In other words, they had obtained the best footage possible – obtained through the hard work of the charity – and now it seemed we had outlived our usefulness and were no longer required. I was gutted. I couldn't believe that they could do such a cold-hearted, calculated thing. I now realised that after all my hard work, time and effort, I had served my purpose and that some faceless person whom I never met in HBO had made a decision without taking any cognisance of the effects that this would have on the charity. I recalled that my only reason for agreeing to make this film happen had been because of the endless possibilities it would give to the charity to raise much-needed funds and to create awareness, especially in the US. But Maryann hadn't finished filming and we already had well-advanced plans for a further trip that October, so what was going to happen now? I expressed how shocked I was about this development and Maryann reassured me that in the editing she would try to ensure that the charity's story still remained. As I had always trusted Maryann, I felt her words were sincere and I agreed to continue doing the planning and organising of the filming schedule required for the October visit.

On my return to Ireland I advised our board of directors of this new twist. The feeling on the board was of dismay that we could be treated in such a casual way and it wanted me to reconsider my position. I told them that my commitment to the film was still there and that I would try and go beyond HBO's decision to renege on the film's primary focus for the greater good of the issues involved. This decision was mine alone and I was acutely aware that I did have a choice and could have walked away from it. But if I had done that, I also realised that the film couldn't go ahead, as Maryann was dependent on us to get back into the country to film, so I agreed to continue. This decision was crucial to how things developed later.

After that second trip, time passed and events overtook us, as we were so busy organising aid programmes, doing speaking tours and raising money. Eventually I got a call from Maryann saying that the film was finished. It had been a mammoth task as she had so much footage, so many stories but so little time. The documentary, which was originally filmed for a one-hour programme, eventually became a 40-minute film.

Then events took on a new twist. Maryann told me that people who had seen it at HBO thought it had the potential to be an Oscar winner and that the programme would be entered right away. Oh my God! I couldn't believe it! This would bring the story to a different level and the potential for publicity and fundraising that could possibly be realised was boundless. After a time, I found myself, along with my assistant, Norrie McGregor, joining Maryann and Angie in LA, where the film would be screened in a public cinema for several nights as part of the nomination process. What an achievement! We were all beside ourselves with excitement. We were joined in LA by members of our American board of directors, Jim and Sherry Douglas and Kathy Ryan.

It was an amazing experience and a very daunting one too. After each screening, Maryann and I would field questions and comments from the audience. Sometimes it felt like we were counselling people, as many were very disturbed and upset by what they had seen. When the screening process ended we stayed on for a few 'chill out' days in LA, as it was a whole new experience, particularly for Norrie and myself. The four of us drove up and down the west coast and saw all the famous places we'd only ever heard of and laughed and joked about all sorts of things. We listened to great Irish rock musicians like Rory Gallagher and felt that life couldn't be better. In our funny moments we played a game of 'what if we won'. We loved to play this game and joked about what we'd wear and who we wanted to see at the Oscars. None of us honestly believed that the film would go any further and felt quite happy as we parted at the airport, congratulating ourselves that we had done great.

Six months and lots of work later, I received a phone call from Sherry and Maryann. I knew the Oscar nominations were about to be announced but wasn't prepared for the teary voice of Maryann saying down the phone, 'We did it! We made it!' I was speechless. For the first time in my life, I couldn't speak, I could only scream! Tears flowed between the three of us across the Atlantic Ocean. The following weeks were weeks of belief and disbelief. What none of us had expected had happened, and we'd be going back to LA. This time, we'd be going to the Oscars Award Ceremony in downtown Hollywood!

All our friends rallied around in Ireland. People sponsored my clothes and a dear friend and fashion designer, Mairéad Whisker, designed and made me a beautiful dress. Friends sponsored flights and before we knew where we were, Norrie and I were back on a flight to LA. On our arrival we went straight to a film festival which was showing the documentary and I arrived just in time to join Maryann for the question and answer session. We went from there to the fanciest hotel we'd ever been in – the Beverly Wiltshire Hotel, where *Pretty Woman* had been filmed. We were overawed. I expected Richard Gere to jump out of our wardrobe any minute! The next day was frantic, but eventually I found myself in the back of a stretch limo with Maryann and her entourage on our way to join the fleet of hundreds of other limos en route to the Kodak Theatre, home of the Oscar ceremony. The journey was an adventure in itself as we passed through all the cordoned-off streets and saw all the snipers on rooftops. Eventually we came closer to the theatre and could see the crowds in the distance. As we got out of our limo we saw all the famous actors and actresses that I had ever admired in my life, and they were right beside and in front of us! The most overwhelming part was going down the red carpet. As I was a total 'nobody', it meant that I could just relax and observe the proceedings. I couldn't really relate to most of it, as it's such a different world to mine, but I did manage to 'people watch' and mentally made a note of everyone I saw and met so I could tell everyone back home.

Eventually the night went on and tension was rising within me as our time drew nearer. It was such a surreal experience that by the time the documentary was announced as the winner I was totally numb. I couldn't articulate my feelings. All I could think of was the people the film was about and what this win could possibly mean for them. I couldn't stay there any longer; I had to get back to Norrie and all our friends who were attending an Oscar party back at the hotel. At this stage we had been joined by the Irish consul, Donal Denham, and his wife, Siobhan, the Irish actress Fionnuala Flannagan and Irish special effects man, Kevin Hannigan. I almost fell into their arms when I got there. The phone starting ringing almost immediately and it was one interview after the other for morning radio in Ireland. It was manic! I got my thoughts together and did dozens of calls over the next 24 hours, but not before I rejoined Maryann and hit the famous Oscar parties. That was an experience! Norrie and I decided to take a quick look around the Vanity Fair party (the most famous one) and then placed ourselves strategically near the entrance and just people watched. We met and spoke with many of our heroes and heroines, such as Susan Sarandon and Tim Robbins.

We eventually fell into bed for about an hour and resumed the radio and newspaper interviews the next day. We were told that the entire island was cheering our success and I kept saying on the radio how there would have been no documentary, not to mention an Oscar, were it not for the stalwart support of the Irish people who had kept the issue alive for 18 years. I also reminded myself and everyone else that the win was a win for the victims and survivors and that our task

was to now translate the Oscar into something of meaning in their lives. They were the real heroes of the day, the people who had allowed us the privilege of entering their lives and telling their stories. Without them we would have had no story, no film.

We were happy to leave LA, I can tell you. It was a world apart for us and it was the closeness of friends that kept us going. We both thought we'd never get back to Cork fast enough! When we eventually arrived home there was an incredible welcome – banners, singing, dancing, cheering, general madness, flowers, children and all our friends and family. The one person I wanted to see was Seán, and there he was on the tarmac, waiting with a big smile and an even bigger hug! I finally relaxed, and only then did I honestly feel we'd succeeded and that it was something powerful, not only for the organisation, but most of all for the people that the story was about.

Slowly life got back to normal and we began to work on the task of making the film a reality for the people of Chernobyl. I honestly felt that if it didn't translate into funding and therefore practical aid, then the Oscar was meaningless in itself, that it was just an empty award. My concern and anxiety rested around that thought and I only felt real relief and hope when Maryann contacted me to say that the broadcast date for HBO to show the film was set for 9 September 2004. The intervening weeks were filled with all sorts of plans between ourselves and our sister organisation in the US. We had to make this broadcast work; we only had one shot at it. If we didn't get money from the 9 September 2004 broadcast, then our chances would be greatly reduced. I was invited by Maryann and HBO to join in one final screening at their studio in New York, where both of us would team up again for questions and answers. This screening was for the American media and they would determine the success or otherwise of the film.

I travelled to New York some days before the broadcast full of expectations that we'd take America by storm and that the whole country would be watching and waiting to respond. Our US sister organisation, in anticipation of massive support, had hired a 400-line telephone response company to deal with the expected onslaught of calls. The first hint that we were being overambitious about the film came at the final media screening at HBO, when very few of the mainstream media were in evidence. It turned out to be mainly a gathering of friends and relations with little or no response from the critics. But, undaunted, we still lived with great expectation and agreed to travel with Maryann and her family to watch the broadcast in West Point Military Academy, as there were a number of students there interested in the film.

Chernobyl Heart was being shown after

another very interesting documentary made by Rory Kennedy, the youngest daughter of Senator Robert Kennedy (younger brother of John F. Kennedy), called *Imagine the Unimaginable*. This was an excellent, hard-hitting documentary looking at the vulnerability of the Indian Point nuclear power plant to possible terrorist attack, so I was convinced that we would attract a very large viewing audience and that, between both films, it would be effective enough to reach for the phone and offer help.

The following evening we got the results of the viewing figures and donations and were very disappointed with both. Our main concern was to get donations in for the charity, but the

With Colin Farrell

With Jim Douglas, Sherrie Douglas, Maryann deLeo and
Kathleen Ryan

With Norrie McGregor

phones remained pretty silent. On the first night we received a mere $270 and since then, having broadcast the film many times, we have received less than $80,000. Our dream of being able to sustain our work through the success of the film wasn't to be realised as we had hoped. This takes me back to the decision taken by HBO to change the focus of the film halfway through. Hindsight is a great thing, but perhaps I should have fought harder for something that would have shown Ireland's magnificent response. That's not my style. I know it was weakness on my part, as I, and the charity, were crucial to finishing the film, so I had some bargaining chips. I had been innocent in my belief that honour and what was right would prevail. I was wrong. What the finished film failed to make clear was what exactly our organisation does, so when people saw the film there was nothing in it to encourage them to respond and give us help.

HBO, a giant of a company, turned out to be very nasty towards a tiny humanitarian organisation. There was no way we could take them on and they knew they held all the cards. Life was made extremely difficult as they refused us the right to show the film for fundraising purposes. This was a severe blow to us, as Maryann had promised us from the outset that we could use it for our own benefit and promotion of our work. Even to get our free phone number up at the very end of a long list of credits was difficult. Again, this was something we had been promised, but now that they had their prize they didn't care about the charity, even less so about those who we had hoped to help. That is what we found most offensive – the fact that the film as a piece of work became the central part of everything and the very people that the film was about were of absolutely no consequence. There were some dreadful phone messages and meetings with some people in HBO who showed no respect, either for those we are trying to help or for any of us in the charity. HBO never contributed any money to our work, which is a very telling fact. But apart from the nastier side of these developments, I still hoped that the film would succeed and bring the issue of Chernobyl back into the spotlight.

However, perhaps all would not be lost, as Maryann was invited to bring the film to Dublin and Cork in Ireland for two major film festivals in early October 2004. We would now have a second bite of the cherry and we put our efforts into getting the best audiences possible together for both screenings. Thanks to our patron, Ali Hewson, we put together a guest list of some of Ireland's most influential people who came together with the Project's key volunteers and made the screening the European Premiere. For the Dublin event we managed to have five of the children present from the Vesnovo

Mental Aslyum featured in the documentary. It was their presence that made the night for all of us, for these are the people who allowed us the privilege of entering their lives and it was out of the story of their misery that the documentary won the Oscar. It was only right and fitting that they were the special dignitaries of the event. Maryann and I were asked to address questions from the audience, but I have to say that Sasha, from Vesnovo, stole the show when he asked to speak. Remembering that until then this boy had little experience beyond the walls of the asylum made his speech all the more poignant. He thanked the people of Ireland as well as Maryann and myself for making the film and he asked that we continue to help the other children left behind who badly needed support. Despite not being allowed by HBO to use the film for fundraising purposes, we still managed to receive over €8,000 from just over 200 people.

The Cork screening for me was a case of 'bringing it all back home', as Cork is the home of the organisation and it's where we planned the making of *Chernobyl Heart*. For me it was the most special screening, particularly as my mother and family were present. More than anyone else, they know how much I'd put into the making of the film and knew the personal price I'd paid for it, and it was very special for us to watch it together at long last. Having Mum present was the icing on the cake, as she had been very ill in hospital and there were doubts as to her ability to recover well enough to attend.

In the meantime, I kept close to reality and made sure the Oscar didn't go straight to my head! After all, this wasn't about winning an award. It was always about the issue and the people affected by the accident. I always kept this in my sights and as a direct follow-through from the film I managed to bring the little boy, Sasha, to Ireland for treatment. He was successfully treated for an unusual form of leprosy, the likes of which had never been seen in Ireland and was mostly associated with the death camps of the Second World War, or the siege of Stalingrad where people contracted it as a result of dirt and bad diet. After a period in isolation he recovered and has become a regular visitor to Ireland. Tatiana, the girl who features in the film when having heart surgery, is now also a frequent visitor to Ireland, where she has reached celebrity status as an 'Oscar winner'.

In 2002 we developed a comprehensive medical care programme for the children of Vesnovo, headed up by one of our stalwart volunteer medics, Johnny Casey, who ensures that out of every four weeks there is a team of Irish medics working with the children for a minimum of eight days. In tandem with the medical care programme, Johnny and his highly skilled medical volunteers do side-by-side training with local nurses. Since the making of the film, we have employed 10 full-time trained nurses, paid for by sponsors from the US and Ireland. This development has helped to ensure that the children are finally receiving the best medical care possible. Along with the 'hands on' medical programme now in place, we have done major renovation work at the asylum. Included in this is the complete rebuilding of a previously collapsed unit. The unit, housing 30 children, was sponsored by our local group in Limerick, Ireland and as the unit was being finished, we flew in a team of builders from our Mallow group to completely overhaul the leaking roof of the entire asylum. So now we can safely say that *Chernobyl Heart* is working for the children, as we have improved their quality of life. We are working our way through the balance of six units, with the end hope that each and every child there will be able to reach their full potential in good living conditions and in a reasonably loving environment.

In October 2004, when completing the documentary Duncan and I had started in 2003, the crew and I had the pleasure of returning to Vesnovo and staying in the new, beautiful Irish unit, complete with clean beds and flushing sit-down toilets. It was heaven! Sam Gracey, our cameraman, filmed the Irish medical team working one-on-one with the children. The difference between 2004 and our first visit was staring us straight in the face! Many of the children who had previously been completely bed-bound were now up and running with the aid of walkers and physiotherapy. The children were now showered and dressed in clean clothes and those who were incontinent were comfortably dry in nappies sent out on our convoy – altogether a different picture to that terrible first visit with Maryann and Angie.

Little did Norrie McGregor and I imagine how much our lives would change after we received that famous SOS fax appeal in January 1991, how immersed and intertwined we would become with the Chernobyl story. What had been a gut reaction to a heart-felt plea has become a major international humanitarian organisation. What had started out as purely a response to children in crisis went on to be a tour de force, developing comprehensive and effective programmes right across the spectrum. What two people unwittingly started in the back bedroom of my house in Cork has grown and become a strong force for real and sustainable change. Thousands of volunteers all over the country have risen to the cry for help and give their time, effort and money, but most of all they give their love. They are the backbone of our organisation. They give helping hands as host families, doctors, nurses, builders, drivers and fundraisers.

The motto of our organisation is 'Offering Hope to Live'. A priest in Belarus said, when he thanked us, that not only did we bring aid but that we brought to the victims 'hope to live', and it struck a deep chord within me and my co-workers, so much so that we adopted it as our organisation's motto, our inspiration.

We have over 60 groups of volunteers located around the whole of the island of Ireland who first and foremost fundraise the whole year round to raise urgent monies and humanitarian supplies for the countless victims of the world's worst nuclear disaster, and secondly bring children from the contaminated zones to Ireland for vital health care, rest and recuperation. Through our 16 aid programmes, our goal is to alleviate suffering and in the process to say very loud and clear, 'Brothers and sisters, your plight and pain is not forgotten.'

As a race I also think we are 'shaped' by those who went before us, many of them missionaries and volunteers for different causes. In our history we know that the courage and strength of many men and women have made us the compassionate people we are today. We have inherited a deep sense of justice, commitment to community and respect for each other regardless of sex, race or class. Our outspokenness on issues sometimes gets us into trouble, but we continue in order to speak for those who feel abandoned, isolated or forgotten. Sometimes these people are within our own society and sometimes from other lands.

It's no accident that we are outspoken for the rights of others internationally who suffer at the hands of injustice, poverty or racism, as we know from our own story that we have suffered in similar ways. The answer to where we get our compassion, activism and volunteerism from lies in our past. In particular, our plight during the Great Irish Famine in the mid-nineteenth century comes to mind. During the darkest times in Irish history, when our

166

own population was practically halved through enforced emigration and mass starvation, the direct hand of solidarity came to us in the guise of shipments of grain and corn to our major ports. We were the recipients of humanitarian aid from the Native American Choctaw tribe and the American Quaker movement. So many of us who are volunteers in the twenty-first century are the descendants of those who survived because of the direct hand of intervention. So when we see the suffering of others, either at home or abroad, it's like an echo of what we have experienced within our own race memory. It's something rooted deep within us.

We know from our own experience that it's only by looking after each other will we be truly 'real' and loving human beings. It's very aptly put in an Irish saying: *'Is ar scáth a cheile a mhaireann na daoine.'* ('It is in the shadow of each other that we truly live.') As an organisation, we believe that it's through working together with others in this field that we not only find 'comfort', but also can be more effective in attaining our goals. To this effect we have formed many alliances and associations throughout the world in Canada, Italy, the US, France and England, to name but a few. However, one joint collaboration will always be particularly special to me, and that is the one we formed with the Israeli organisation Chabad's Children of Chernobyl and its director, Jay Litvin. Jay met Helen

Faughnan and I at a UN conference on Chernobyl in Moscow a number of years ago, and we instantly liked each other. Over the following years we joined together at many events and made our voice stronger through sharing thoughts and ideas. Jay was dedicated to the victims and had a brilliant mind, which he used through the power of his writing and vocal skills. He worked extensively in the Chernobyl region of Ukraine and to a lesser extent in Belarus, and believed that it was as a result of his work in the zones that he contracted cancer. He battled bravely and serenely as his body was slowly taken over by the rampaging disease. Jay died in 2004, and with his passing we have lost one of Chernobyl's greatest advocates. May he rest in peace.

For the first 15 years of our work, we concentrated our efforts and resources on projects designed to give immediate assistance in the form of emergency relief. Our efforts focused on ameliorating the condition of those most affected by the disaster by the direct provision of medical and humanitarian assistance. This involved the development and implementation of a wide range of programmes, including biannual aid convoys that have delivered €65 million ($80 million) in direct and indirect aid, a long-term care programme that has offered terminally ill children access to medical treatment otherwise beyond their reach and several orphanage refurbishment projects which have dramatically improved the living conditions of children in state-run institutions. In addition, we operate an annual emergency airlift of children from the heart of the radiation zones to Ireland for rest and recuperation. To date we have brought over 14,000 children to the safe Irish environment. We also have an ambulance programme which has delivered and maintains a fleet of over 160 ambulances for hospitals, institutions and orphanages throughout the region, and we run long-term community development projects.

Along with direct humanitarian intervention programmes, we also work closely with the scientific community in Belarus on radio-ecological education and implementation. We fund the work of Professor Vasily Nesterenko of the Belarusian Institute of Radiation Safety through the purchase of Mobile Radiation Monitoring units which carry teams of scientists and their equipment throughout the contaminated zones checking the levels of radiation in both children and foodstuffs. The testing units are designed by Ukrainian and German scientists as a direct result of the accident and show the practical application of technology. They allow us to identify the children most at risk and in need of recuperation and medical treatment in Ireland. As over 500,000 children still live in the zones, the radiation monitoring units are vital in providing basic radiation protection.

These programmes have made real and tangible differences to the lives of countless people.

While the humanitarian and emergency relief aspects of our work will continue to be a large part of our work as long as a need for such traditional interventions remains, we have also identified the need to build capacity in other directions, directions that emphasise and give priority to the long-term developmental approach.

In 2001 we did a lot of soul-searching about our work, looking in particular at the area where we prop up state asylums and orphanages. We started to question the whole culture of the abandonment of children, and in our search for answers we looked at international best practice. The trend worldwide is now to close institutions and reintegrate the residents, thus breaking away from what could be called a type of apartheid. In the long term, the plight of the children held in institutions, rather than being helped, is exacerbated. Half a century of research has shown that no matter how much you improve the institutions, they are still detrimental to the mental, emotional and physical wellbeing of the children. In other words, children thrive best in a loving home environment placed in a wider community. We must believe in the families and trust in their love for their children. We also had to look at the issue of improving conditions in institutions.

In trying to get to the root of this culture of abandoning children, the answers were heartbreakingly clear. It's because of fear of the future, deep despair, a sense of pervasive hopelessness, isolation and abject poverty. Out of this we have chosen a whole new approach that has moved us away from the old-style donor versus recipient model to one that is all about partnership, giving the people a say in how they want their dying communities to come to life again. We now have partnerships with several fledgling indigenous organisations that work in and among local communities throughout the contaminated regions. This work has convinced us that our efforts and resources are best channelled into areas that promote and foster a culture of self-help among the people.

This approach is one that is increasingly championed by a number of international organisations. It's about finding long-term recovery through new initiatives designed to assist the individuals and the communities to take their future in their own hands by fostering a process of healing through measures that will extend healthy life and improve wellbeing. In practice, this means that a strong emphasis must be put on efforts to improve household incomes, to build and strengthen primary health care and help in the process of rebuilding the structures of society at local community level. The best way to achieve these objectives is via strong collaboration between government agencies, both national and international, the voluntary sector and the communities at local level.

As poverty and unemployment continue to plague the lives of people still living in the zones as well as the evacuees, it's crucial that massive investment is given in order to break out of the poverty and dependency cycle, thus stimulating economic development at local level by establishing small and medium-sized enterprises in the affected areas which will promote self-sufficiency. This would have to include training and support of such income-generating activity. Such local initiatives need to be geared towards small-scale household self-sufficiency that requires only small-scale financing. For example, the promotion and setting up of local credit unions would be an ideal way of making this kind of initiative become a reality and would encourage local recovery strategies. This, along with the encouragement of small-scale worker and consumer co-operatives, would be a great step forward. This would require building local community leadership and help to build a new generation of proactive initiatives that would be holistic and integrate primary health care, the environment and economic measures as part of the plan for the future. Resources need to be focused on the most affected communities and individuals, identifying families on low income who grow their own food, creative ways of educating such families about how to live safely in the zones and showing specific methods of growing food that would be less dangerous to eat.

We had started to put this theory into practice with the help of the Irish Department of Foreign Affairs and Pfizer International, who have given us matching funds for our first community building project. It started in 2003 when we identified our new strategy in moving from the emergency to recovery period. We visited Belarus in April 2003 with a view to working with the community in Zhytkovitchy, where we had identified a very good local leader and supporters within the fledgling day care centre movement. We had been seeking a strong local community base for some time, and after two years of research we were back in the local government building negotiating an agreement between ourselves and the authorities to rebuild a 100-year-old building as a new day care centre. It was a brave step on our part, as we had never entered into such a relationship in Belarus. It was a bold building plan, designed by architect Duncan Stewart. I say 'bold' because it was the first building that we know of in the region that was to be built to the highest Western standards, and as we are committed to energy efficiency, this building has the latest, state-of-the-art insulation and a wood gasification heating system integrated with 16 square metres of solar energy collectors placed in the new roof. The collectors provide all the centre's hot water needs and excess energy collected is used to supplement the surface heating requirements for eight to nine months of the year.

This new centre, as planned, was to become the hub of the town and surrounding villages by providing a wide range of services. It would provide short-term refuge for at-risk children, skills workshops for the disabled young adults, home help service for the retired and disabled elderly community, counselling services for women in crisis, a citizens' advice centre, literacy and computer classes, skills educations and much more. It would cater for 17,000 people spread out around 140 villages. It would give support and back-up for the most marginalised of society, and in the long term be an alternative to state-run orphanages and asylums, thus allowing families to look at options other than abandoning the children they couldn't cope with.

It was an inspiring project right from the very beginning, headed up by volunteer Project manager Ryan Cleary of Pfizer Ireland and supported by Cork-based Project manager Simon Walsh. Over the period of two years from signing the contract, Ryan and Simon drew up a work schedule that would involve Ryan travelling to the region three times a year, bringing with him each time a team of 20 volunteer builders. Side by side with our highly skilled builders, we employed six local tradesmen. The Belarusian workers were also given work in between team

visits in order to complete the project on time, as well as fulfilling our commitment to local partnership through employment. I visited the building site on several occasions and at times was overcome by the absolute dedication, bordering on devotion, of all the workers, both Irish and Belarusian, to the project. They worked all the hours from dawn until dusk, in a freezing cold dusty shell of a building, and yet each and every one of them smiled, laughed and sang their way through the job. Throughout their time there they lived in extremely basic living conditions, but the camaraderie was always high. I was more than impressed, and from the positive outcome and success of the new day care centre, opened in May 2005 by the Irish ambassador Justin Harmon, we plan to replicate it in other villages throughout the zones.

Keeping all our aid programmes financially secure and afloat is a huge challenge for us, as we have no regular source of income, but we have a regular and ever-growing expenditure. Continuity and security of programmes is vital to our effectiveness, but we don't have a 'product' or 'commodity' to sell, so we don't have any way of generating income in the business sense. And yet we are a business of sorts. We are professionally run by a wonderful team of staff and we have a workforce of 20 doctors, nurses, physiotherapists and tradespeople in the region. In planning sustainable, long-term recovery projects, we need to constantly work three to five years in advance in terms of planning and projections. But this becomes precarious without any income, so fundraising takes a massive bite out of both Ali's and my time. Over the years we have been lucky and for a long time we have been sustained by the small fundraising activities of our committed volunteer workforce. For example, our groups have funded our entire fleet of 160 ambulances and the airfares for the 14,000 children to come here to Ireland for rest and recuperation. But we realised early on in the life of the charity that we would need much more funding to sustain the quality and calibre of our work on a more permanent basis. To this effect we have initiated a number of high-profile events to fund our work. In 1996 Simon, Rory, Tony, Andrew and Rebecca Coveney from Cork sailed around the world in a 50ft boat, the *Golden Apple*. Their extraordinary two-year, 30,000 mile journey was the single biggest fundraiser we have ever had. U2 gave us the gift of their song 'Sweetest Thing' and turn up at many events to lend their support. Ali and her friend Caroline Downey organised very successful international fashion shows, bringing supermodels to support both our Project and the Irish Society for the Prevention of Cruelty to Children (ISPCC). They managed to bring in Naomi Campbell, Kate Moss, Christy Turlington, Helena Christensen and Jerry Hall, to name but a few! Along with the models came some of the world's best-known musicians, such as Mick Jagger. The world of high fashion and rock music is totally alien to my world, and I often felt a bit of an outsider at these events, but it was great knowing that these people had been touched by our work and felt concerned enough to give us a helping hand.

Apart from the support of the music and modelling industries, we have worked very hard to bring the business and corporate world on board. Luckily for us, we found two businessmen in particular who have been crucial in our much-needed requirements for regular income support. Michael Roden, a Dublin-based businessman, saw some of our moving footage on television and he immediately took action and became the generous sponsor of our office premises. Up to that point the entire organisation and staff worked out of my home, which meant that Seán and I had practically little or no private life. So when Michael offered to rent us premises, not only did he give the organisation its own 'home', but he also gave Seán and myself our privacy again. Since then Michael and his wife, Helen, have supported us financially for many different projects. Then another man, Noel Kelly, a self-made and kind businessman who has many companies, ranging from marketing to graphics and events management, contacted me and offered his help. Noel has provided us with annual fundraisers, including a ball, and getting prominent Irish artists to donate paintings for auction. Both Michael and Noel provide the charity with vital

income, but the moral and personal support they provide me is of equal value. I know that whenever I need advice they are the ones I can turn to.

The challenge for Ali and I as the public faces of the organisation is to constantly come up with fundraising ideas. Every New Year I feel slightly anxious when I know that for the following 12 months one of my tasks will be to raise another €1.5 million. But with friends like Ali, Michael, Noel and others, I know the money will come in. I have never yet been let down by those who have promised to help. I feel humbled by what others do for us as well as being truly grateful. I call Ali my guardian angel; she has been there for me all these years, minding me when I'm tired, standing me up when I'm down and wiping away the tears of exhaustion and frustration that nobody else sees. Ali and I are a great combination, as she is far more strategic, calm and pragmatic, and I'm the one who's the ball of fire eager to plunge in! Ali's reticence and my spontaneity give us a good balance and it's because of Ali's forward thinking that we launched our American sister organisation. This development is for the long-term financial support of a variety of US-based projects in the Chernobyl region co-run from both sides of the Atlantic.

Our hope for our future is that we continue to stabilise financially through the encouragement of people to fund our work right across the spectrum, from our local groups throughout Ireland to the business and corporate community of both Ireland and the US, thus enabling us to continue doing what we do best.

But our future isn't just about financial security. It's about something much deeper and more profound. It's about the substance and equality of life and how we add our contribution to its essential essence. I'm thinking particularly now of the Irish families who have gone as far as adopting the Chernobyl children of the fallout and making them dual citizens of Ireland and Belarus. I'm thinking, for example, of Nastiya (Anna) Gabriel, Alexei Barrett, Raisa Carolan, Philippe Fitzmaurice and Vanya Cadogan, all of whom come from the same 'No 1' Home for Abandoned Babies. These were the first 'pioneers' of our fragile adoption agreement. I have a very close relationship with both Anna and Raisa and every time we meet I feel a welling up of emotion of such joy, pride and absolute love for them. Both of the girls have such shining personalities, giving an extraordinary peace and joy to everyone they meet. They seem to live each and every day as if it were their last. Through them we learn how to appreciate life anew.

Despite our optimism, we became deeply concerned about our future in Belarus following a speech made in mid-November 2004 by President Lukashenko. Part of his speech spoke out strongly against foreign adoption when he said, 'It is a shame for the state.' A ban was placed on all foreign adoptions. He went on to criticise the effects on children of rest and recuperation travel programmes abroad. His speech also contained strong and negative opinions on international aid programmes. On consulting with a number of experts on the true meaning and possible impact of the speech, I learned that the pattern of Belarusian presidential speeches is that they tend to pave the way for opinions expressed which later became law, so we had a clear indication that it was the intention of the Belarusian government to possibly restrict the movement of children going abroad for rest and recuperation and for treatment.

Even if a complete or partial ban wasn't going to be enforced, we were deeply concerned that more restrictive measures would be put in place that would prohibit certain groups of children from travelling. For example, we had heard that children with haemophilia or epilepsy would no longer be allowed to travel abroad. It also became apparent that cumbersome procedures were being put in place that would make convoys of aid very difficult.

As one of the only Chernobyl NGOs with United Nations NGO status, we felt that it was only right and proper that we head up the international campaign to highlight these developments and work through diplomatic initiatives to get the president to reconsider his position. This,

174

in tandem with our Minister for Foreign Affairs leading the political response, meant that we really had a chance. Thanks to Minister Dermot Ahern's swift reaction, diplomatic meetings were held between our ambassadors and their Belarusian counterparts in Moscow, Warsaw and London. We were able to clearly say to the president of Belarus that we would not be prepared to passively accept any law that would jeopardise the work of our organisation. While we realised President Lukashenko's frustrations with the West because of the sanctions imposed by both the EU and the US, we consistently reiterated that we were above and beyond politics, that this was a humanitarian issue. Children's lives and health were at risk.

Luckily, I had planned to have a meeting with our own president, Professor Mary McAleese, on 25 November and I decided in advance that she should be briefed on what was happening in Belarus. I needed the full weight of the organisation behind me and phoned Ali, who was in New York with her husband Bono to launch their latest album, and asked her to fly back to Ireland that night to arrive just in time for the president's meeting in Dublin. True to her commitment, Ali flew in and we rushed to Áras an Uachtarain (the home of our president), where we were welcomed with open arms. When we laid out the picture of events, President McAleese showed deep concern and

support and we discussed various options of how, as President of Ireland, she might quietly intervene on behalf of the children and ourselves. I came away from the meeting with a much lighter heart, as I knew that we had the backing of our president which, along with the backing of our government, had to stand for something.

Our voice was heard and we were guaranteed by the Belarusian officials that our children coming in that Christmas would be allowed to travel as planned and that we weren't to worry about the next groups of 1,200 children due to travel to Ireland in 2005. Through diplomatic channels the Belarusians went to great lengths to reassure us that nothing would change between our two countries. I have had high-level meetings with many Belarusian diplomats and politicians who have advised me that nothing will hinder the work of NGOs in Belarus. I have great faith in people's word and I believe that their message is a sincere one. Our own lives would be worthless if we were to give up and abandon these innocent victims. Thanks to the diplomacy of internal and external parties, the ban on foreign adoptions was lifted in 2005.

It's a privilege to lend a helping hand, a real gift. A Jewish proverb says it beautifully: 'One heart is mirror to another.'

When Maryann deLeo was asking the composer Henryk Górecki for permission to use part of his Third Symphony as the background closing music to the Oscar-winning film, he said that permission would be granted 'only if the film offered hope'. Can we offer these innocent victims hope? I mean not just kind platitudes, but real hope, hope that translates into positive action, that will renew their tired and battered spirits, give hope to countless parents who watch their precious children suffer so needlessly, offer hope to struggle on into an uncertain future, hope to conceive and bear healthy children, hope to believe that their struggle isn't ignored, hope to believe that they aren't on their own. Can we offer them hope to hope? On this twentieth anniversary, let us remember the catastrophe that is Chernobyl; that suffering, that sacrifice; and now, that neglect, that negation of reality. Children are every nation's treasure and true wealth; they are our special investment in the future; they are the nation's bequest to future generations.

By reaching out to the children, we are sharing the burden, breaking the isolation, doing something. I emphasise the 'doing something' because fine words will never feed a hungry child or heal a broken body. The action of the 'doing' translates all the positive inner energy into that blanket, that food, that medicine, that bed, into that warm hug that really saves the world.

Chernobyl has resulted in the worldwide realisation that the earth is our common home, a beautiful grain of life in the depths of the universe, having become so fragile in the hands of man.

The twentieth anniversary of the Chernobyl disaster offers us that special opportunity to reflect again on what happened to other innocent victims and their children and children's children. Unlike other disasters, a nuclear disaster carries on and on and on, wreaking havoc in generation after generation, twisting and distorting the very web of life. We can't ignore the latest generation of Chernobyl victims and we mustn't abandon the future generations of victims. We must remind ourselves that we are human beings who can show compassion, continue to do what we can and to support them in every way we can.

Having spent much of October and November 2005 in Ukraine and Belarus listening to and observing filmmakers and journalists ask the same questions time after time, I'm convinced that they are asking the wrong questions. They ask, 'How many people died? How many will die? Is this or that cancer or illness definitely caused by radiation? What is Chernobyl? How much radiation were you exposed to? Why do you all look so healthy? Show me the evidence!' These are questions with non-specific answers or answers that don't satisfy the required neat

EPILOGUE: BEACON OF HOP

'The word that God has written on the brow of every man is hope.'
Victor Hugo

176

logic. We seek absolutes in a situation where there can be no absolutes, no definitive answers, for we ask the wrong questions. Expectations of seeing something grotesque and distorted distract from the true effects, with no realisation that any dose of radiation is an overdose. This puts the burden of proof of radiation-related injuries on the victims.

If we continue to seek only logical, rational answers, we will constantly be diverted from the true picture, a picture of human and ecological fragility showing us how delicately balanced the relationship between man and nature is; a picture of

how precarious life has become in the hands of man.

I now believe that as long as we try to place Chernobyl within our existing understanding of catastrophes, understanding it will continue to elude us. Our experiences from other disasters is clearly inadequate because we are facing a realm of the unknown not previously experienced. It requires a new understanding, a new bravery, a new kind of courage – an approach with an open mind allowing us to see things in a way that will place the value of human life at its core.

Pope John XXIII said, 'The accumulation of vast wealth while so many are languishing in misery is a grave transgression of God's law, with the consequence that the greedy, avaricious man is never at ease in his mind: he is in fact a most unhappy creature.' It's true – we are being offered the wonderful privilege of saving the life and soul of another human being.

I dedicate these closing words to all my family, friends, colleagues, co-workers, volunteers and supporters who have travelled many a road on this journey with me, inspiring me and encouraging me in all my efforts. I thank every one of you and congratulate you for your love and humanity, not just towards me, but more

Giving away your time, your energy, your money, your concern for the sake of another person, someone you don't know, in a land that you never travelled to – that indeed is true love, true Christianity in practice, and the return profit of peace and happiness in your life is beyond your wildest dreams.

so for the love and compassion and humanity you have shown towards the dear innocent victims of Chernobyl, especially to the children and their families.

I congratulate and celebrate all the other Chernobyl support groups that are also working long and hard throughout the world. Let us continue to journey together and let us rededicate our lives and energies to continuing with this journey of love and sharing and caring for our dear brothers and sisters whose lives have been blighted by the Chernobyl disaster, 20 long, hard years ago, on 26 April 1986.

I thank and salute all the wonderful people of Belarus, western Russia and northern Ukraine. Their struggle is exemplary and long. Their hearts are broken but their spirit is strong. God bless you all!

Christmas 2005
Cork
Ireland

The explanations and definitions given here are sometimes simplified or incomplete. More rigorous definitions can be found in technical books.

absorbed dose: The amount of radiation energy absorbed in a unit of mass of an organ or living thing from ionising radiation.

alpha particle: An electrically charged positive (+) particle emitted from the nucleus of some radioactive chemicals, e.g. plutonium. It contains two protons and two neutrons. It's the hardest of the atomic particles emitted by radioactive chemicals and can cause ionisation. Alpha particles are relatively heavy and large and cannot easily penetrate clothing or skin, but they are highly dangerous if inhaled or ingested. See also 'ionisation'.

atom: The smallest particle of matter that can take part in a chemical reaction. Some atoms may combine to form larger, but still tiny, molecules.

background radiation: This includes emissions from radioactive chemicals which occur naturally and those which result from the nuclear fission process. The meaning of this term is vague. In a licensing process it includes radiation from all sources other than the particular nuclear facility being licensed, even if the source includes a second nuclear facility located on the same site (US regulation). Radioactive chemicals released from a nuclear power plant are called 'background' after one year.

beta emitter: A nuclide which releases beta radiation.

beta particle: An electrically charged negative (–) particle emitted from some radioactive chemicals. It has the mass of an electron. Krypton-85, emitted from nuclear power plants, is a strong beta emitter. Beta particles can penetrate a centimetre or more of living tissue but can be stopped by thin sheets of denser materials. They can also cause ionisation. See also 'ionisation'.

Chernobyl AIDS: Radiation-induced AIDS which breaks down the immune system.

dose: Energy imparted to matter by nuclear transformations (radioactivity). This is the general term for the quantity of radiations absorbed by the body.

dosemeter (sometimes spelled 'dosimeter'): A device used to measure doses of ionising radiation.

caesium-137: Is produced from the detonation of nuclear weapons and emissions from nuclear power plants. Highly toxic and highly explosive in cold water. Chemically, it resembles potassium.

electron: A negatively charged atomic particle that is lighter than protons and

181

neutrons. All atoms are made up in part of electrons. A free electron is one which isn't associated with a nucleus.

element: A substance which consists only of atoms of the same atomic number and which cannot be split up into simpler substances by chemical means. There are over 100 different elements, of which 92 occur naturally.

fallout: Radioactive materials which descend from the atmosphere onto the surface of the earth. Fallout may be caused by nuclear explosions, by the release of radioactive substances after a nuclear accident and by the routine operation of nuclear facilities.

gamma rays: Electromagnetic radiation of high energy and penetration released by some nuclear transformations. They are similar to X-rays and will penetrate through the human body. Iodine-131 emits gamma rays. Both gamma rays and X-rays cause ionisation. See also 'ionisation'.

gene: A unit of heredity composed of DNA, occupying a fixed position on a chromosome. A gene may determine a characteristic of an individual or regulate or repress the operation of other genes.

genetic effects of radiation: Those effects which are experienced by the children of the individual receiving the radiation or by later generations – in contrast to somatic effects, which occur in the individual receiving the radiation, and teratogenic effects, which occur in the embryo or foetus inside the mother's womb.

graphite: A soft form of carbon used as a moderator in nuclear reactors.

half-life, biological: The time required for the body to eliminate one half of an administered dose of a radioactive chemical. It's influenced by health and diet. The main methods of elimination are via urine, faeces, exhalation and perspiration.

half-life, physical or radioactive: The time taken for the activity of something radioactive to lose half its value by decay. The chemical resulting from the transformation may be either radioactive or non-radioactive. The process of radioactive decay is independent of temperature, pressure or chemical condition. Half-lives range from less than a millionth of a second to millions of years. The half-life is a characteristic constant for each particular nuclide. An individual nuclide may decay before or after the half-life. Radiation is released every time a radioactive material changes to the next material in its decay series.

Radioactivity per unit weight is inversely proportional to the half-life. For example, a specified quantity of caesium-137 (half-life 30.04 years) is about 76,000 times more radioactive than the same quantity of caesium-135 (half-life 3 million years). Iodine-131 is a radioisotope with a half-life of eight days.

In traditional nuclear physics there is a rule of thumb that after 10 half-lives a substance is considered to have decayed to a 'safe' level. However, this rule does not consider the size of the original quantity of the radioactive nuclide, nor is there a universally accepted definition of what is 'safe'.

hot spot: A local geographic area where radioactive environmental contamination levels are higher than the average for the surrounding area. Hot spots may be a tiny point containing only a few radioactive fragments, or they may encompass many square kilometres, such as when an air mass or cloud containing a high concentration of radioactive dust becomes part of rain or snow. Areas of body tissue where a much higher than average concentration of radioactivity occurs are also referred to as hot spots.

ionisation: Radiation with enough energy to ionise atoms, i.e. to remove electrons from them. It includes X-rays and gamma rays and the particles that are emitted from radioactive substances. High levels of ionising radiation can have noticeable, severe and lethal effects on the human body. At low levels, none of our senses can detect ionising radiation. We cannot see it, hear it, feel it or smell it. However,

although we are unable to detect it with our senses, damage to the human body can occur which only becomes apparent years later. Radioactive nuclides pose the greatest threat to human health when they are inhaled or ingested. However, radiation-emitting fragments can be so small that they fasten in the many sweat pores and hair follicles all over the body.

There are three main types of ionising radiation: alpha, beta and gamma. Alpha emitters are the most harmful to living cells if ingested or inhaled, but the distance which alpha particles can travel is only a few centimetres in air and mere tens of microns in soft tissue. Beta particles can travel up to about 80 cm in air and they cannot go through steel or wood thicker than about 1 cm.

The difference between alpha and beta particles is like that between cannonballs and bullets. Alpha particles, like cannonballs, have less penetrating power but more impact. For this reason the biological damage of alpha radiation is considered to be about 20 times that of the same absorbed dose of beta or gamma radiation.

Gamma rays are a type of electromagnetic radiation, unlike alpha and beta radiation, which are forms of particle radiation. Gamma radiation can be as harmful as beta radiation and can travel great distances. The majority of gamma radiation is stopped by a few centimetres of concrete.

When discussing the distance which different forms of radiation are capable of travelling, it's important to distinguish between radioactive particles and radioactive emitters. Alpha and beta particles travel a very short distance, due to radioactive decay. In contrast, minute dust fragments containing alpha, beta and gamma emitters can be transported great distances by wind and water. Sometimes the false impression is given that locations only centimetres away from a radioactive source are isolated from exposure. See also 'alpha particle', 'beta particle' and 'gamma rays'.

isotope: One of two or more forms of an element that have differing atomic weights, i.e. that have the same number of protons but different numbers of neutrons. It's considered that after 10 half-lives a radioactive isotope has decayed into another radioactive isotope (a daughter product) or has become stable. See also 'radioactivity'.

low-level radiation: This includes background radiation and man-made radiation from low-level nuclear waste. The International Commission of Radiological Protection assesses radiation damage on the premise that radiation is always potentially harmful, however small the dose.

natural background radiation: Low-level radiation from space and from radioisotopes in rock, soil and building materials. It includes emissions from radioactive chemicals which aren't man-made. These chemicals include uranium, radon, potassium and other trace elements. They are made more hazardous through human activities such as mining and milling, since this makes them more available for uptake in food, air and water.

neutron: A neutral elementary particle. In the nucleus of an atom it's stable, but when free it decays.

plutonium: A man-made, highly toxic synthetic metallic element with a half-life of 24,400 years. Traces of plutonium occur in uranium ore and it's produced in a nuclear reactor by neutron bombardment of uranium-238. The isotope plutonium-239 readily undergoes fission and is used as a reactor fuel in nuclear power stations and in nuclear weapons.

Plutonium is an extremely dangerous substance because of its radioactivity and the fact that when ingested as an oxide or other compound it deposits in the bone and is excreted very slowly. Metallic plutonium isn't absorbed by digestive organs. Inhalation of only a few thousandths of a gram may lead to death within a few years and much smaller quantities can cause lung cancer after a latent period of about 20 years.

Plutonium should be handled by remote control using extreme caution to avoid the release of dusts to the atmosphere. Plutonium metal is highly reactive and thus must be stored at low temperatures in dry air to avoid corrosion.

It was first identified in experiments at the University of California in 1940, and plutonium-239 was isolated a year later. There are 16 isotopes of plutonium, of which only five are produced in significant quantities: plutonium-238, -239, -240, -241 and -242.

proton: A stable, positively charged particle in an atomic nucleus.

rad: The old unit of absorbed dose of ionising radiation. One rad is equal to 1 rem for most forms of radiation. The rad was first defined in 1953. In 1956 the rad replaced the roentgen (a unit of radiation exposure) for clinical work involving X-rays or radioactive sources. In practice the rad and the roentgen both represent about the same amount of energy, since 1,000 rad is applicable to all types of radiation. Historically, the rad can be traced back to 1918, when it was suggested as the name for the unit of X-ray dose sufficient to kill a mouse.

radiation: The emission and propagation of energy through space or tissue in the form of waves. This term usually refers to electromagnetic radiation, classified by its frequency: radio, infrared, visible, ultraviolet, X-rays, gamma rays and cosmic rays.

radioactivity: The emission of radiation from atomic nuclei. Nuclear radiation includes alpha and beta particles, gamma rays, X-rays and neutrons. See also 'isotope'.

rem: The acronym for roentgen-equivalent-man. This unit of dose equivalent was replaced by the sievert.

somatic effects of radiation: Those effects which are experienced by the individual receiving the radiation – in contrast to genetic effects, which occur in offspring or future generations, and teratogenic effects, which occur in the embryo or foetus inside the mother's womb.

strontium-90: A radioisotope with a half-life of 28.2 years. It's one of the hazards of fallout.

teratogenic effects of radiation: Those effects which are experienced by the embryo or foetus inside the mother's womb – in contrast to genetic effects, which occur in children of the individual receiving the radiation or in later generations, and somatic effects, which occur in the individual receiving the radiation.

uranium: The heaviest naturally occurring element. This dark grey, radioactive, metallic element was discovered by the German chemist H.M. Klaproth in 1798. Uranium is both radiologically and chemically toxic and poses a health hazard as a heavy metal as well as a radioisotope. Uranium-235 is used as a source of nuclear energy by fission.

Isotope	Radiation	Half-Life	Parts of the body affected
americium-241	alpha	433 years	blood, bones
barium-140	beta & gamma	13 days	bones, reproductive organs
caesium-137	beta & gamma	30 years	muscles, reproductive organs
carbon-14	beta	5,600 years	bones
cobalt-60	beta & gamma	5 years	liver
cobalt-60	gamma	5 years	reproductive organs
iodine-131	beta & gamma	8 days	reproductive organs, thyroid
krypton-85	beta & gamma	10 years	reproductive organs
krypton-95	beta & gamma	10 years	lungs
phosphorus-32	beta	14 days	bones
plutonium-239	alpha	24,400 years	blood, bones, lungs, reproductive organs
plutonium-240	alpha	6,600 years	blood, bones
polonium-210	alpha	138 days	spleen
potassium-42	beta & gamma	12 hours	muscles, reproductive organs
promethium-137	beta	2 years	bones
radium-266	alpha	1,620 years	bones
radon-222	alpha	3.8 days	lungs
ruthenium-106	beta & gamma	1 year	reproductive organs
strontium-90	beta	28 years	bones

GUIDE TO RADIOISOTOPES

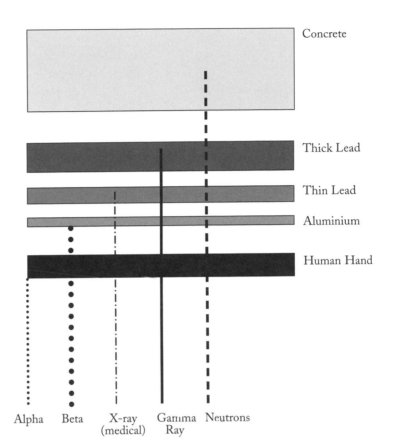

187

Bass, Corin Fairburn and Kenny, Janet (eds.), *Beyond Chernobyl: Women Respond*, Sydney, NSW: Envirobook, 1993.

Bertell, Rosalie, *No Immediate Danger: Prognosis for a Radioactive Earth*, London: Women's Press, 1985.

CORE (Cumbrians Opposed to a Radioactive Environment), www.corecumbria.co.uk.

WISE (World Information Service on Energy), Netherlands, www.nirs.org.

World Health Organization, Reactor Accident Report, 6 May 1986.

SOURCES